WILDLAND FIRE DYNAMICS

Wildland fires are among the most complicated environmental phenomena to model. Fire behavior models are commonly used to predict the direction and rate of spread of wildland fires based on fire history, fuel, and environmental conditions; however, more sophisticated computational fluid dynamic models are now being developed. The quantitative analysis of fire as a fluid dynamic phenomenon embedded in a highly turbulent flow is beginning to reveal the combined interactions of the vegetative structure, combustion-driven convective effects, and atmospheric boundary layer processes. This book provides an overview of the developments in modeling wildland fire dynamics and the key dynamical processes involved. Mathematical and dynamical principles are presented, and the complex phenomena that arise in wildland fire are discussed. Providing a state-of-the-art survey, it is a useful reference for scientists, researchers, and graduate students interested in wildland fire behavior from a broad range of fields.

KEVIN SPEER is a professor and Director of the Geophysical Fluid Dynamics Institute at Florida State University, with experience in field and laboratory measurements of turbulent geophysical flows. He recently developed a new program in Fire Dynamics at Florida State University that combines the fields and faculty of numerous departments, including Earth, Ocean, and Atmospheric Sciences; Mathematics; Scientific Computing; Statistics; Physics; and Engineering.

SCOTT GOODRICK is a research meteorologist with the US Forest Service Southern Research Station and serves as Director of the Station's Center for Forest Health and Disturbance in Athens, GA. He has been working as a research scientist with the Forest Service, specializing in fire–atmosphere interactions and smoke management for 18 years. Prior to joining the US Forest Service, he spent four years as the fire weather meteorologist for the state of Florida, helping to develop their Fire Management Information System.

WILDLAND FIRE DYNAMICS

Fire Effects and Behavior from a Fluid Dynamics Perspective

Edited by

KEVIN SPEER
Florida State University

SCOTT GOODRICK
US Forest Service

<CAMBRIDGE
UNIVERSITY PRESS

CAMBRIDGE
UNIVERSITY PRESS

University Printing House, Cambridge CB2 8BS, United Kingdom

One Liberty Plaza, 20th Floor, New York, NY 10006, USA

477 Williamstown Road, Port Melbourne, VIC 3207, Australia

314–321, 3rd Floor, Plot 3, Splendor Forum, Jasola District Centre, New Delhi – 110025, India

103 Penang Road, #05–06/07, Visioncrest Commercial, Singapore 238467

Cambridge University Press is part of the University of Cambridge.

It furthers the University's mission by disseminating knowledge in the pursuit of education, learning, and research at the highest international levels of excellence.

www.cambridge.org
Information on this title: www.cambridge.org/9781108498555
DOI: 10.1017/9781108683241

© Kevin Speer and Scott Goodrick 2022

First published 2022

A catalogue record for this publication is available from the British Library.

ISBN 978-1-108-49855-5 Hardback

Additional resources for this publication at www.cambridge.org/9781108498555

Cambridge University Press has no responsibility for the persistence or accuracy of URLs for external or third-party internet websites referred to in this publication and does not guarantee that any content on such websites is, or will remain, accurate or appropriate.

Contents

Colour Plates section to be found between pp. 128 and 129

Contributors

Mohamed Ahmed
Department of Fire Protection Engineering, University of Maryland, Baltimore, MD

Rachel L. Badlan
School of Science, University of New South Wales, Canberra

Bret W. Butler
USDA Forest Service, Rocky Mountain Research Station, Missoula, MT

Janice Coen
National Center for Atmospheric Research, Boulder, CO

Miguel Cruz
CSIRO Land and Water Flagship, Canberra

Jean-Luc Dupuy
French National Institute for Agriculture, Food, and Environment (INRAE), Paris

James E. Hilton
Data61, CSIRO, Clayton South, Victoria

Nigel Berkeley Kaye
Clemson University, Clemson, SC

Rodman Linn
Los Alamos National Laboratory, Los Alamos, NM

Richard H. D. McRae
Australian Capital Territory Emergency Services Agency, Canberra

Joseph J. O'Brien
USDA Forest Service, Southern Research Station, Athens, GA

François Pimont
French National Institute for Agriculture, Food, and Environment (INRAE), Paris

Susan Prichard
University of Washington School of Environmental and Forest Sciences, Seattle, WA

Daniel Rosales-Giron
Tall Timbers Research Station, Tallahassee, FL

Eric Rowell
Division of Atmospheric Sciences, Desert Research Institute, Reno, NV

Jason J. Sharples
University of New South Wales at the Australian Defence Force Academy, Canberra

Timothy M. Shearman
Tall Timbers Research Station, Tallahassee, FL

Kevin Speer
Geophysical Fluid Dynamics Institute, Florida State University, Tallahassee, FL

Andrew L. Sullivan
Data61, CSIRO, Clayton South, Victoria

Christopher M. Thomas
Climate Change Research Centre, University of New South Wales, Sydney

Ali Tohidi
Department of Mechanical and Aerospace Engineering, San Jose State University, San Jose, CA

Arnaud Trouvé
Department of Fire Protection Engineering, University of Maryland, Baltimore, MD

J. Morgan Varner
Tall Timbers Research Station, Tallahassee, FL

Salman Verma
Department of Fire Protection Engineering, University of Maryland, College Park, MD

1

Wildland Fire Combustion Dynamics

The Intersection of Combustion Chemistry and Fluid Dynamics

ANDREW L. SULLIVAN

1.1 Introduction

An uncontrolled high intensity wildfire is one of the most terrifying natural phenomenon anyone may have the misfortune to experience firsthand. The sheer terror generated by the immense energy released in the combustion zone of a wildfire, a terror that may recall primordial subconscious fears of the absolute uncontrolled power of Mother Nature in what may seem to be a highly chaotic and unpredictable manner, can have lasting effects upon those impacted by the fire.

The term "spread like wildfire" is part of the general vernacular of our society but even then the meaning of such a phrase is not consistent, much like the behavior of a wildfire in its nature. Meanings of this phrase include "uncontrolled,"[1] "to be or become known,"[2] "to spread or circulate or propagate very quickly and widely, to spread rapidly,"[3] and "to quickly reach or affect a lot of people."[4] In all these, there is the sense of something moving rapidly and uncontrollably, yet haphazardly but incessantly.

A wildfire, then, is something considered to be uncontrollable, unpredictable, and unstoppable. Except that we know that this is not necessarily the case. Where the perception may be of a roaring inferno propagating unstoppably with immense energy and intensity, simultaneously there will be sections of the same fire quiescent and mild, hardly spreading at all. Where there may be the observation of a fire front consuming great amounts of biomass and converting it to constituent molecules and atoms to be released into the atmosphere, other sections of the same fire may barely singe the lowest layers of the vegetation. The perception of wildfire is not

---

[1] Cambridge Dictionary: https://dictionary.cambridge.org/dictionary/english/spread-like-wildfire (last accessed November 18, 2021).
[2] Macmillan Dictionary: www.macmillandictionary.com/dictionary/british/spread-like-wildfire (last accessed November 18, 2021).
[3] The Free Dictionary: https://idioms.thefreedictionary.com/spread+like+wildfire (last accessed November 18, 2021).
[4] Collins Dictionary: www.collinsdictionary.com/dictionary/english/spread-like-wildfire (last accessed November 18, 2021).

necessarily the whole story of fire and without a full and complete understanding of fire there can be no hope of ever being able to predict it or control it.

Noted French philosopher, Gaston Bachelard (1884–1962), in his work *The Psychoanalysis of Fire* (Bachelard 1938, p. 2), observed:

We are going to study a problem that no one has managed to approach objectively, one in which the initial charm of the object is so strong that it still has the power to warp the minds of the clearest thinkers and to keep bringing them back to the poetic fold in which dreams replace thought and poems conceal theorems. This problem is the psychological problem posed by our convictions about fire … Fire is no longer a reality for science. Fire, that striking immediate object, that object which imposes itself as a first choice ahead of many other phenomena, no longer offers any perspective for scientific investigation … The reason for this is that the question [what is fire?] has fallen within a zone that is only partially objective, a zone in which personal intuitions and scientific experiments are intermingled. As a matter of fact, we shall demonstrate that our intuitions of fire – more perhaps than any other phenomenon – are heavily charged with fallacies from the past. These intuitions lead us to form immediate convictions about a problem which really should be solved by strict measurement and experimentation.

It may be said that these observations remain as true today as they were when first published. Much of the operational knowledge of wildland fire in use today (be it wild or otherwise) stems from the *ad hoc* learnings gained from long experience and direct observation of fire in the landscape. For many decades this empirical font has enabled land managers and fire bosses to successfully manage fire in the landscape. But as the well of knowledge grows older and retires, as the climate and conditions in which fires burn changes, the ability to transition traditional heuristic systems into more structured and applicable paradigms quickly hits limits and intuitions and rules of thumb begin to fail, with the potential for deleterious and catastrophic outcomes. Very often, however, the admirable desire for fire and emergency managers to utilize "evidence-based" learnings for "transparent" decision making becomes adulterated by the need to urgently fill knowledge gaps or implement "novel" solutions without critical and robust analysis of the veracity of such responses. As Bachelard observed, the lack of objectivity in relation to fire at every level means that beliefs and intuitions (or worse) are quite often treated equal to (or better than) the results of strict scientific investigation.

In this chapter we will explore the intersection of two fields of study that is at the very heart of the behavior of wildland fire but for which the domains are of two completely separate scientific disciplines. These are the chemistry of biomass fuel, combustion, and heat release, and the physics of fluid dynamics and heat transfer. It is within this zone that, like Bachelard remarks, fire "is only partially objective, in which personal intuitions and scientific experiments are intermingled." For, while the study of the problem of fire can and has traditionally been isolated within each discipline, a complete understanding of fire cannot be achieved without the other.

The problem may be summarized in one very simple, and common, question: When starting a fire, why does blowing on the nascent embers help? As we will see, the intuitive answer of "by providing more oxygen" is not completely correct, as the preliminary reactions necessary for combustion to initiate do not involve oxygen.

We will begin by looking at the chemical composition of biomass through which wildland fires propagate, and the chemical processes and reactions by which fuels thermally degrade and react with the atmosphere to liberate heat and energy. We will then consider the processes in which these reaction pathways are influenced by the environment around the combustion zone of a fire and how the combustion zone can in turn influence the environment around it. In the nexus of the two domains we shall consider observations of fire behavior that can only be understood by joint application of each domain of knowledge.

A conceptual model of the processes involved in the propagation of a wildland fire was developed by Sullivan (2017a), who attempted to capture the key processes involved, including sources and sinks of energy and closure pathways, in the ignition, combustion, and spread of fire in wildland fuels. The key processes identified in this model may be reduced to a very simple consideration of the cycle of heating, degradation (often mistakenly called "pyrolysis"), and combustion of biomass fuel (Figure 1.1).

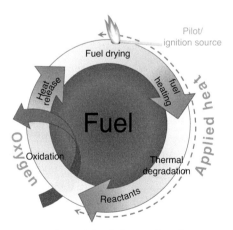

Figure 1.1 Simple conceptual model of the cycle of processes involved in the combustion and spread of wildland fire. If a fire starts with a pilot flame or ignition source and sufficient heat is applied to biomass fuel, the fuel will undergo evaporation of free moisture and dehydration that results in hydrolyzed cellulose that under continued application of heat begins to thermally degrade and produce pyrolysis and char products (reactants) which then oxidize in flaming and glowing combustion that release heat. If that heat is sufficient and localized enough, sustained spread of fire may occur.

This simplification, of what is a very complex set of chemical and physical mechanisms, identifies a closed loop inside of which the concept of sustained fire spread exists. This loop begins with the introduction of a pilot flame or ignition source to fuel which then undergoes evaporation of moisture and dehydration to produce a modified fuel structure called *activated cellulose* which physically appears very similar to the original fuel. Continued application of heat from the ignition source or pilot then begins to thermally degrade the activated cellulose which produces pyrolysis and char products (or reactants, since the reaction processes have barely begun at this point). These reactants will then energetically oxidize in the presence of air as flaming or glowing combustion, releasing heat and combustion products in the form of smoke and other emissions.

It is the stage of oxidation that the primary effect of the combustion environment comes into play, primarily in the form of oxygen in the air surrounding the combustion zone. The introduction of air – and the motion of that air – introduces turbulence (and also generates turbulence through the interaction of the energy released in combustion and the presence of the air) that affects the efficiency of the oxidation processes as well as the efficiency with which heat is transferred to adjacent fuel that enables the fire to spread. Thus, it is both the chemistry of the combustion of the fuel and also the physics of the flow of air and transfer of heat in and around the combustion zone that determines whether the resulting fire is a raging inferno or a quiescent smolder.

1.2 Combustion Chemistry

1.2.1 Chemistry of Wildland Biomass

Wildland fuel is composed of the live and dead biomass elements that make up the finer components of the vegetation. While the majority of the biomass in a wildland setting is held in the larger elements, such as tree boles, stems, and branches, these are generally too large to easily ignite and contribute meaningfully to the behavior and spread of a wildland fire (McArthur 1967; Rothermel 1972). Thus, it is the finer biomass elements, generally <6 mm in diameter, that provide the primary source of energy driving the behavior of a wildland fire front (McArthur 1967; Rothermel 1972). It is changes in the way these fuels combust (particularly in regard to efficiency of energy transfer; Anderson and Rothermel 1965) that affect the behavior and spread of the fire front.

In most wildland settings, the fine biomass fuels consist predominantly of dead fallen leaf litter, bark, twigs, shrubs, and grasses (Beall and Eickner 1970). When antecedent and prevailing weather conditions are severe, the fuel can also include larger material such as fallen branches, intermediate and overstorey tree canopies, and even standing trees (Gould et al. 2011) but, due to their size, these generally do not contribute significantly to the dynamics of the fire front.

Live and dead biomass fuel, therefore, represents a wide range of physical structures, plant components, age, and level of accumulation and decomposition, each of which depends on the type of wildland setting and history of the land which can influence the inherent flammability of a fuel (Varner et al. 2015; Grootemaat et al. 2019). The primary constituents of biomass fuel are cellulose, hemicelluloses, and lignins, the distributions of which vary considerably across plant species and plant parts. Indicative ratios for biomass across a broad range of species and functional components (e.g. leaf, twig, stem) are 25–50% cellulose, 15–39% hemicelluloses, 12–44% lignins, and 0–33% other components (consisting of minerals, water, and extractives) (Sullivan 2017a).

Cellulose is a naturally-occurring long-chain, linear, unbranched polysaccharide ($[C_6H_{10}O_5]_n$) of β-D-glucose monomers ($C_6H_{12}O_6$) (Shafizadeh 1982; Williams 1982) in $\beta(1, 4)$ linkage (Figure 1.2). This polymer is a nonreducing carbohydrate and ranges in length from 200 to 10,000 units with molecular weights of 250,000–1,000,000 or more (Morrison and Boyd 1983, p. 1113). It is the most abundant organic material on our planet and is found in the protective walls of plant cells, particularly in stalks, stems, trunks, and all woody portions of plant tissues (O'Sullivan 1997). Cellulose is also present in bacteria, algae, fungi, and some animals, and plays a very important role in the human diet as fiber. Cellulose

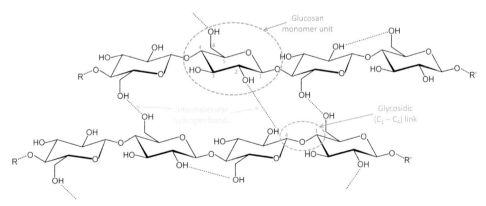

Figure 1.2 Example of the skeletal formula of a portion of two adjacent cellulose chains, indicating some of the intra- and inter-molecular hydrogen bonds (dotted lines) that may stabilize the crystalline form of cellulose. R and R' indicate continuation of the cellulose chain being reducing and nonreducing, respectively. A glucosan monomer with carbon atom numbering convention is shown by the dashed ellipse in the top chain. The glycosidic link between C_1 and C_4 of two adjacent glucosan units is shown by the dashed ellipse in the bottom chain. Reprinted with permission from Springer Nature Customer Service Centre GmbH: Springer Nature, Current Forestry Reports (Sullivan 2017a) © 2017

is essentially the same in all types of biomass, differing only in the degree of polymerization (O'Sullivan 1997).

The glycosidic bond in cellulose is between C-1 of one β-D-glucose residue and the hydroxyl group on C-4 of the next unit (β(1, 4) linkage). The bond is formed through the process of condensation or dehydration between two glucosan units (i.e. a water molecule is formed and released in the joining of two D-glucose residues in this manner) as part of the larger process of photosynthesis within a plant.

The natural cellulose polymer is a straight chain with no coiling and adopts an extended rod-like morphology (O'Sullivan 1997). Parallel chains of cellulose can form hydrogen bonds in which surplus electron density on hydroxyl group oxygens is distributed to hydrogens with partial positive charge on hydroxyl groups of adjacent chains (see Figure 1.2). Multiple parallel chains may thus bond to form a crystalline structure that is very rigid and forms the basis of the microfibrils of plant cells, where bundles of up to 1,000 cellulose chains are bonded in parallel, and contribute to its high tensile strength (Jane 1956). Segments of naturally occurring cellulose can exhibit regions of both crystalline structure with ordered alignment of both inter- and intra-molecular bonds and amorphous structure in which the bonding is disordered (Broido et al. 1973) but not entirely random (O'Sullivan 1997). The ratio of crystalline to amorphous structures is in the order of 70:30, with the crystalline regions being relatively unreactive compared to those of amorphous regions (Ball et al. 1999a).

As a result of its rigid structure, cellulose is extraordinarily stable. It is insoluble in water, relatively resistant to acid and base hydrolysis, and inaccessible to all hydrolytic enzymes except those from a few biological sources. As a result, cellulosic fuels can take a long time to biologically decompose and require considerable energy to thermally degrade. Unlike starch, which has a crystalline-to-amorphous transition (i.e. breakdown of interchain hydrogen bonds) at 330–340 K in water, it takes ~ 590 K at a pressure of 25 MPa for cellulose to become amorphous in water (Deguchi et al. 2006).

Hemicelluloses are complex polysaccharides (generally copolymers of glucosan and a variety of other possible, mainly sugar, monomers) that occur in association with cellulose. Hemicelluloses generally consist of branched structures comprising 50–200 monomeric units and a few simple sugar residues, but their structures vary substantially depending on the biomass species and functional component. These polysaccharides are soluble in dilute alkaline solutions (Yaman 2004). The most abundant hemicelluloses are xylans, which are found in cell walls and made from monomers of xylose, a pentose sugar.

Lignins are highly branched aromatic polymers consisting of phenylpropane monomers in varying concentrations, depending upon the species, cell type, and functional component. As with hemicelluloses, they are generally found in cell walls, especially in woody species, and are often bound with cellulose to form a lignocellulose complex (Yaman 2004). Lignins are the second most abundant biopolymer after cellulose and they provide rigidity and physical strength to plants (Gordobil et al. 2016).

A large variety of other elements and compounds are found in biomass fuels. These include starches (i.e. nonpolymer carbohydrates), minerals and trace elements (such as nitrogen, phosphorus, potassium, calcium, magnesium, sodium, and silica amongst others), water, and salts (Demirbaş 2004). Very minor amounts of metals, such as mercury, may also be found (Howard et al. 2019). Extractives and inorganics including terpenes (isoprene polymers) and resins (fats, fatty acids, and fatty alcohols) may also be found in fuels. After complete combustion, some of these, particularly minerals, appear as residual ash.

1.2.2 Chemistry of Wildland Fuel Combustion

Cellulose is the most widely studied substance in biomass combustion, with less focus being given to the study of the thermal degradation and combustion of hemicelluloses or lignin (Di Blasi 1998), perhaps as a result of the relative thermal instability of these compounds. Although the extrapolation of the thermal behavior of any particular individual component to describe the kinetics of a chemically complex fuel such as biomass is only an approximation (Di Blasi 1998), the properties of cellulose have been found to exert a dominating influence on the rates of thermal degradation of biomass. As a result, the detailed thermokinetics of cellulose provides the best understanding of the combustion of biomass fuels (Williams 1982). The remainder of this discussion focuses on the thermokinetic properties of cellulose; however, differences in the purity and physical properties of the cellulose, hemicellulose, and lignin in the biomass can also play an important role in the degradation process (Di Blasi 1998), with inorganic matter acting as a possible catalyst or an inhibitor of the degradation reactions.

1.2.2.1 Combustion Processes

While the word "combustion" is generally used to describe any high-temperature, self-sustaining oxidation reaction (Babrauskas 2003, p. 14), it is also used to describe the complete process in the conversion of unburnt fuel to ash and burnt residue (Luke and McArthur 1978). The combustion of biomass in wildland fire is not strictly a linear sequence of events, one in which fuels are preheated, they ignite, distilled gases combust, and then finally residual char is combusted (as portrayed, for example, in Gisborne 1948; Vines 1981; Pyne et al. 1996). Although there are several unique stages identifiable during the propagation of a flame front, these are not necessarily sequential and can sometimes occur simultaneously or in competition with each other.

The first step in biomass combustion is heating of the fuel, often initially from a pilot source such as a direct flame or spark but also by the transfer of heat via radiation, convection, or conduction from an approaching flame front. Under continued heating, free and bound water on and in the fuel evaporates or is liberated. If sufficient heating occurs, the fuel then begins to thermally degrade and its structure changes fundamentally and irreversibly. The primary products of this thermal degradation are combustible gases (i.e. volatiles) and char. These gas and solid phase products then oxidize in air, in flaming combustion (gas phase) or glowing and smoldering combustion (solid phase).

It is often believed that a fuel must reach a specific temperature for ignition to occur. In truth, however, while the temperature of the reactants is important in determining initiation of combustion, ignition does not occur at a single fixed temperature but the reactions that may be perceived to be "ignition" become more prevalent at higher temperatures.

All chemical reactions require reactant molecules to be brought together in the correct orientation with sufficient kinetic energy to break or form bonds between or within the reactants (Morrison and Boyd 1983, p. 55). The minimum kinetic energy for a reaction to occur is called the *activation energy* and the rate at which that reaction proceeds is proportional to the concentration of the reactants and the rate of collisions between reactants that occur at or above the activation energy and in a favorable orientation, which can be expressed as the empirical relation known as the Arrhenius equation (Moore 1963, p. 273):

$$r = A \exp^{\left(\frac{-E_a}{RT}\right)}, \tag{1.1}$$

where A is a pre-exponential factor usually considered to be a constant for any particular reaction that subsumes the molecule collision and orientation factors and represents the frequency of collisions in the correct orientation (s^{-1}), E_a is the reaction activation energy (J mol^{-1}), R is the universal gas constant (8.314 J K^{-1} mol^{-1}), and T is the temperature of the reactants (K). This equation reveals the important role of reaction temperature in the rate constant through the exponential dependence, with a small increase in temperature resulting in a large increase in the rate constant. Values for A and E_a are generally derived from thermogravimetric analysis obtained by measuring changes in mass and overall system energy while a sample is being heated at a fixed rate under an inert atmosphere (Antal and Várhegyi 1995; Antal et al. 1998) and may be correlated (Philibert 2006).

Reaction enthalpy, ΔH_R (kJ mol^{-1}), is the change in enthalpy when a reactant forms a product following a reaction: $\Delta H_R = H_f(products) - H_f(reactants)$, where H_f is the standard state heat of formation of the reactant or product. When ΔH_R is positive, the process absorbs heat (i.e. it is endothermic) and, when it is negative, the process releases heat (exothermic).

The reactions involved in combustion of biomass fuel generally have such a relatively high activation energy and, thus, are highly temperature sensitive, that when a fuel is heated it undergoes a long incubation period during which it appears to change relatively little. When combustion finally initiates, it does so in dramatic fashion over a very small (<50 K) temperature range (Atreya 1998), giving the impression that a single ignition temperature exists.

The rate of the formation and oxidation of volatiles is much faster than that of the formation and oxidation of char and thus appears to occur first in the passage of the flame front, with the char oxidation appearing to occur after the fire front has passed. In a wildland fire, all reactions can occur simultaneously and some reactions occur at the expense of others (i.e. they are competitive). It is the dynamic nature of these reactions that can result in the complex behaviors observed in bushfires.

1.2.2.2 Thermal Degradation

Under thermal stress, cellulose undergoes thermal degradation (or thermal decomposition) reactions that commence in the range 400–500 K in which reactants that liberate the bulk of the energy during combustion are formed. In cellulose, thermal degradation occurs along two pathways: *char formation* and *volatilization*. Each of these pathways involves the fundamental modification of the underlying cellulose structure with different activation energies and promoting conditions and are fundamentally competitive in that only one or the other pathway can occur within a given section of chain. As a result, different reaction products with different enthalpies and reaction rates can be produced with each having a very different impact on overall behavior of a fire.

Char formation or *charring* occurs when cellulose fuel (F) undergoes inter- and intra-chain cross-linking and dehydration under thermal stress (Broido and Weinstein 1970; Weinstein and Broido 1970). This is more likely to occur in crystalline regions than amorphous regions (Ball et al. 1999a) and can occur via two distinct processes. Primary charring is a slow chemical conversion of a fuel that, due to a low activation energy, begins at relatively low temperatures (Eq. (1.2)). Continued heating of the fuel causes cross-linking of the carbon skeleton of the structure resulting in the elimination of water (dehydration), carbon monoxide (decarbonylation), and carbon dioxide (decarboxylation), and the formation of the desaturated anhydrous carbohydrate commonly known as char (C). In this process, the underlying morphology of the original fuel is retained as the internal structure of the substrate is maintained by the cross-linking reactions (Mok and Antal 1983).

This reaction path has an activation energy of $\simeq 110 \, kJ \, mol^{-1}$ (Di Blasi 1993; Diebold 1994), a collision frequency of $6.7 \times 10^5 \, s^{-1}$ and a reaction enthalpy of $\simeq -1$ to $-2 \, kJ \, g^{-1}$ (Milosavljevic et al. 1996; Ball et al. 1999a).

$$ F \xrightarrow{heat} Char + CO + CO_2 + H_2O + heat. \tag{1.2} $$

Secondary charring occurs at a higher temperature as a result of a higher activation energy in the presence of moisture (Eq. (1.3)). It occurs in competition with the volatilization reaction pathway following thermal scission (or thermolysis) of the cellulose chain at a C_1–C_4 glycosidic link and involves rehydration of the cellulose chain via intermolecular nucleophilic action. Here, the competitive nature of reactions means that only one or the other can occur; they cannot both occur. In this case the nucleophile that bonds to the thermolyzed carbocation at C_1 is a water molecule which may be present within the fuel substrate or the result of previous dehydration reactions occurring via primary charring (Ball et al. 1999a). The initial product is a reducing end which has "lost" the opportunity to volatilize, known as *hydrolyzed cellulose* (OH). The released hydrogen ion from the water molecule bonds to the negative ion forming a nonreducing end that can undergo no further reactions. Similar to the formation of char via the primary pathway, continued heating of the fuel causes crosslinking of the carbon skeleton of the structure with dehydration, decarbonylation, and decarboxylation desaturating the cellulose chain resulting in an anhydrous char species. While the initial glycosidic thermolysis is endothermic, the subsequent crosslinking reactions result in a net exothermicity. This reaction path has an activation energy of $\simeq 200 \, kJ \, mol^{-1}$ (Di Blasi 1993; Diebold 1994), a collision frequency of $6.9 \times 10^{22} \, s^{-1}$, and a reaction enthalpy of $\simeq -1$ to $-2 \, kJ \, g^{-1}$ (Milosavljevic et al. 1996; Ball et al. 1999a).

$$ F + H_2O \xrightarrow{heat} OH \xrightarrow{heat} Char + CO + CO_2 + H_2O + heat. \tag{1.3} $$

While the morphology of char formed through secondary charring is often less structured than that of primary char as a result of the thermolysis of the cellulose chains, the crosslinking reactions maintain some semblance of the original fuel substrate. The degree of molecular desaturation determines the darkness of the charred fuel substrate (Coblentz 1905).

Volatilization is a reaction that occurs in competition with that of the secondary charring pathway following thermolysis and involves intramolecular nucleophilic attack of the resonance-stabilized positive center on C_1 through donation of the electron density on the C_6 hydroxyl oxygen (Ball et al. 1999b, 2004) (Eq. (1.4)). It generally occurs in conditions of low or nil moisture and comprises cyclization, depolymerization, and the release of a levoglucosan (1,6-anhydro-β-D-glucopyranose or $C_6H_{10}O_5$) molecule via thermolysis at the next C_1–C_4 glycosidic linkage in the chain (Ball et al. 2004).

$$F \xrightarrow{heat} LG \xrightarrow{heat} V. \qquad (1.4)$$

This reaction is endothermic, requiring about 300 J g^{-1} (Ball et al. 1999a), has a collision frequency of 2.8×10^{19} s^{-1}, and an activation energy of about 240 kJ mol^{-1} (Di Blasi 1998). Levoglucosan is a solid at ambient temperatures, often described as a "tar" (Williams 1982), but is gaseous at the temperature of thermal degradation, with a melting temperature of 455–457 K. As a result of its volatility, it is the source of a wide range of species following further thermal degradation that readily oxidize in secondary reactions, resulting in what is seen as flaming combustion. Wodley (1971) identified nearly 40 products from the thermal decomposition of levoglucosan – many of which were products of reactions between initial volatiles – including pentane, acetalaldehyde, furan, furfural, in addition to 20 others previously identified, including formaldehyde and formic acid.

Thermolysis of the crystalline anhydrous cellulose tends to occur near the ends of the chains rather than at any arbitrary glycosidic link elsewhere as they are more accessible to the nucleophile (Ball et al. 1999a). This results in scission of glucose monomers, cellobiose, and short oligomers. The key morphological difference between the two competing thermal degradation pathways is that fuel that has undergone the volatilization process does not retain any of the original fuel's structure and becomes amorphous.

The critical element involved in the combustion of biomass fuel is the competitive asymmetric chemistry between two nucleophiles competing for the positively charged carbon center on C_1 of the carbonium ion formed in thermolysis of a glycosidic bond in the cellulose chain. The nucleophiles are either a molecule of water or an -OH group on C_6 on the glucosyl-end of the cleaved chain. Hydrolyzed cellulose (OH), a reducing chain fragment with the tendency to undergo the crosslinking reactions (i.e. dehydration, decarbonylation, and decarboxylation), then results if the water molecule is successful. Under continued heating this will produce charcoal. If the -OH group is successful, a levoglucosan-end that is resilient to the crosslinking reactions and that depolymerizes to levoglucosan, known as *levoglucosan-end cellulose* (LG), will form, which will lead to the formation of other volatile products.

Broido (1976) determined from experimental weight loss data that under moderate heating cellulose undergoes an "incubation" period before branching into either the depolymerization or char formation reactions. He concluded that, although there was no weight loss during this period, the fuel underwent important changes that dictated which subsequent path would dominate under further heating (i.e. the nucleophile competition). The product of the incubation period has been called *activated cellulose* and correlates to the formation of hydrolyzed cellulose or levoglucosan-end cellulose.

1.2.2.3 Smoldering and Glowing Combustion

Oxidation of char appears in the form of smoldering or glowing combustion and involves solid phase (fuel) and gas phase (oxidant) components reacting across the interface of the two phases. As a result, there is not a large range of intermediate pathways that can be interrupted by turbulent mixing of reactants. However, due to the many possible char species with a large range of carbon:hydrogen ratios, the possible reaction pathways can be similarly numerous.

Activation energies for char oxidation vary significantly, with higher carbon concentration species having much higher activation energies than more saturated forms. Oxidation of pure carbon requires temperatures in the order of 1,000 K, whereas more saturated species with less double bonds require 700 K (Harris 1999). While char is normally quite refractory to high temperatures, recently formed hot char can oxidize easily (Sullivan and Ball 2012).

As an example, if we assume a sample char species that has undergone some dehydration and decarboxylation (e.g. $C_{15}H_6O$), then the basic stoichiometric reaction for the solid phase oxidation would be:

$$C_{15}H_6O + 16O_2 \rightarrow 15CO_2 + 3H_2O + \text{heat.} \qquad (1.5)$$

This reaction can lead to a range of intermediate species such as CO or even elemental carbon if the reaction stops and oxidation is not complete, such as when reactants are cooled below the reaction's activation energy or oxygen is limited. As this reaction occurs across the interface between the solid phase and gas phase, it is possible for gas phase oxygen to be restricted or even excluded from reaching the solid phase reaction surface by the presence of reaction products such as ash which may build up during the reaction and which may form an insulating layer over the reacting surface, halting the reaction. In contrast to the oxidation of volatiles (Eq. (1.6)), it can be seen that, for this particular example char species, more than 2.5 times the amount of oxygen is required for the complete oxidation of one mole of fuel, which alludes to the sensitivity of this solid-phase oxidation reaction to the availability of oxygen for completion.

When solid phase reactions continue to completion, there remains a characteristic fine white ash combustion residue composed mainly of minerals, salts, and other inorganic components of the fuel that do not combust and which contains very little carbon (Surawski et al. 2016).

Oxidation of char is highly exothermic ($\Delta H \simeq -32 \text{ kJ g}^{-1}$), over twice that of the volatile oxidation, has a lower activation energy $\left(E_a \simeq 180 \text{ kJ mol}^{-1}\right)$ but occurs at a much slower rate $\left(A \simeq 1.4 \times 10^{11}\ \text{s}^{-1}\right)$ (Eghlimi et al. 1999; Branca and di Blasi 2004).

Due to the slow reaction rate of both the formation and oxidation of the char, much of the char oxidation appears to occur after the passage of the flaming front (Surawski et al. 2015) and often involves larger fuel particles, leading to the impression that glowing or smoldering combustion only occurs after flaming combustion has ceased. Much of the heat released is confined to the fuel bed, with little transported away from the combustion zone.

Solid-phase oxidation reactions of char occur in two regimes, leading to two modes of char combustion. These regimes are kinetic-controlled and diffusion-controlled reactions (Williams 1977). Kinetic-controlled reactions occur where the concentration of oxygen is not limited and the rate of reaction and heat release is strongly dependent on the temperature of the reaction. Diffusion-controlled oxidation occurs where the oxygen concentration at the reacting char surface is small compared to the ambient oxygen concentration (i.e. combustion is oxygen limited) and the rate of heat release is governed by the rate of diffusion of oxygen to the surface of the fuel.

Kinetic-controlled char oxidation leads to the char combustion mode described as "smoldering" and typically yields less complete oxidation with greater amounts of particulate or partially combusted emissions and much slower propagation rates (Ohlemiller 1985), often to the point of extinguishment (Boonmee and Quintiere 2005). Diffusion-controlled char oxidation leads to the char combustion mode described as "glowing" and generally results in more complete consumption and less particulate emissions with higher surface temperatures. Boonmee and Quintiere (2005) identified a critical temperature of 670 K that defined the two regimes; below this value smoldering would dominate, above this value glowing would dominate.

The primary difference between smoldering and glowing char combustion modes is the degree to which the underlying oxidation regime determines the completeness of the oxidation reactions. Smoldering combustion generally produces larger amounts of partially combusted particulate emissions, most often seen as smoke, where the oxidation reaction has ceased before complete oxidation of the fuel. Glowing combustion on the other hand produces few visible indicators with the exception of the dull red glow where sufficient oxygen is present at the burning fuel surface which results in complete oxidation of the char. However, in daylight this regime is not easily detected with the naked eye, being saturated by other light sources such as the sun or flames (for examples see figures in Section 1.4).

Char oxidation, whether smoldering or glowing, is the most difficult type of combustion to extinguish, particularly since water is a key ingredient in the char formation pathway, which complicates suppression efforts. Effective suppression can require very large amounts of water (Rein 2013); as much as one to two liters of water per kilogram of burning fuel is required to stop combustion (Rein 2016).

1.2.2.4 Flaming Combustion

Oxidation of the volatilized gas-phase levoglucosan and its derivatives occurs in what we see as flaming combustion. These reactions are highly complex and disordered due to both the chemistry involved and the susceptibility of the reactions to turbulence in the oxidant and fuel flows affecting the mixing of reactants in what is known as *turbulent diffusion* (Bilger 1989). Studies of emissions of the turbulent diffusion flames from the combustion of bushfire fuel (such as those of Wodley 1971; Hurst et al. 1994; Greenberg et al. 2006) show that the number of oxidation products is considerable and often the result of many incomplete intermediate reaction paths. At its simplest, however, the stoichiometric reaction for levoglucosan oxidation can be written as:

$$C_6H_{10}O_5 + 6O_2 \rightarrow 6CO_2 + 5H_2O + \text{heat.} \tag{1.6}$$

This form of the reaction assumes that intermediate reactions are complete, but the number of pathways that such reactions can take is quite large and not all paths will result in complete combustion to carbon dioxide and water. Bowman (1975) identified 30 possible pathways for the combustion of methane (CH_4), an example of one of the many intermediates of the thermal degradation of levoglucosan. Intermediate species included CH_3, H_2CO, HCO, CO, OH, and H_2. Elemental carbon often forms in these reactions, particularly when combustion is incomplete, and appears as soot. It is the visible radiation from heated particles of carbon that we see as flames (Gaydon and Wolfhard 1960).

These reactions may be further complicated by the presence of other elements such as nitrogen in the atmosphere, which can lead to the formation of a variety of nitrogen oxide (NO_X) species as well as toxins such as dioxins and polycyclic aromatic hydrocarbons (Gullett and Touati 2003; Lemieux et al. 2004) that can be extremely harmful, especially to firefighters (Reisen et al. 2006).

These gas-phase oxidation reactions are highly exothermic and very fast. Oxidation of levoglucosan has an activation energy of approximately 190 kJ mol^{-1}, occurs with a collision frequency of about 2.55×10^{13} s^{-1}, and has a reaction enthalpy, ΔH, of approximately -14 kJ g^{-1} (Parker and LeVan 1989).

Turbulent diffusion oxidation occurs when the hot combustible gases from volatilization (fuel) mix diffusively with oxygen in the cooler ambient air (oxidant) in a highly turbulent environment (Bilger 1989). These react where conditions (e.g. mixing ratio, temperature) are suitable at the interface between the fuel and oxidant (Vervisch and Poinsot 1998). This relatively thin interface is called the *reaction zone* and surrounds the envelope of volatile gas as it mixes and reacts with the oxygen in the air. As a result, the oxidation reactions are often mixing-limited (that is, the oxidation reaction does not occur because the fuel and oxidant cannot be brought together in the correct stoichiometric ratio) and can result in large volumes of volatiles that separate from the reaction zone before oxidation has commenced or has completed. These gases may disperse or may subsequently burn out some height above the fire when it mixes with oxygen if it remains hot enough (Cheney and Sullivan 2008).

At any stage in the oxidation process, any reaction pathway may stop (through loss of energy or reactants) and its products advected away from the reaction zone (Sullivan 2017b). These partially combusted components form the bulk of what appears as smoke and other emissions (Andreae and Merlet 2001). The more turbulent the reaction zone, the more likely that reactants will be removed before complete combustion, hence the darker and thicker smoke from an intense section of flame front fire, as opposed to the lighter, thinner smoke from less turbulent flames.

1.3 Combustion Processes and Environmental Interactions

The foregoing material presents the current understanding of the chemical processes and pathways involved in the combustion of cellulosic biomass fuels. However, these chemical reactions do not occur in isolation or in a vacuum. By their very nature they are contained within the environment in and through which a wildland fire propagates. As a result, the conditions that define the initial and boundaries of these reactions are not constant and very often are not able to be known in great detail. Much of the knowledge thus far gathered about the thermal degradation of cellulosic biomass and subsequent oxidation of thermal degradation products comes from laboratory settings where such conditions are strictly controlled and experimental quantities are subject to negligible uncertainty. In this section we will look at how the environmental conditions in which biomass combustion occurs influences those reactions and results in what we see as wildland fire. Furthermore, the physical evidence for these processes and pathways can be found in careful observation of free-burning fires and we will investigate examples from free-burning laboratory and field scale wildland fire experiments for the evidence of such chemistry.

1.3.1 Competitive Thermokinetics

The nucleophilic competition between the volatilization pathway and the charring pathway in the thermal degradation of cellulosic biomass may be an important component of much of the apparent capricious behavior witnessed in free-burning

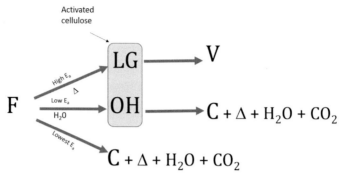

Figure 1.3 A simple schematic illustrating the macro-scale effect of the thermal degradation of cellulose fuel substrate (F) resulting in the direct formation of char at low temperatures (F → C), the nucleophilic competition resulting in either levoglucosan-end cellulose (LG) and the formation of volatiles (V) through the path (F → LG → V) or hydrolyzed cellulose (OH) and the formation of char (C) at high temperatures through the path (F → OH → C). Δ indicates heat. Levoglucosan-end cellulose and hydrolyzed cellulose may be described as activated cellulose since there is no significant mass loss during this thermal degradation stage.

biomass fires, particularly the change in fire behavior around the perimeter of a fire under relatively constant conditions. The outcome of the competitive thermokinetics of this reaction is critically dependent upon thermal and chemical feedback during the reactions, which are both influenced by the macro-scale environment within which the reactions take place and influence the macro-scale fire behavior.

Figure 1.3 is a simple schematic illustrating the macro-scale effect of the thermal degradation of cellulose derived from that of Ball et al. (1999a). Thermal degradation of a fuel substrate (F) may result in the direct formation of char (C) at low temperatures. At higher temperatures thermal degradation may result in the formation of activated cellulose in the form of either levoglucosan-end cellulose (LG) or hydrolyzed cellulose (OH), depending on the outcome of the nucleophilic competition, with no significant loss of mass at this stage. Under continued heating, the levoglucosan-end will form volatiles (V), whereas hydrolyzed cellulose will form char. The exothermic char-forming reactions release heat as well as water and gases such as carbon dioxide as a result of the crosslinking reactions. In contrast, the volatile-forming reactions are endothermic with no by-products.

1.3.2 Chemical and Thermal Feedbacks

Figure 1.4 illustrates the chemical and thermal feedback involved in the thermal degradation reactions of cellulose and it is here that the complexity of the reaction

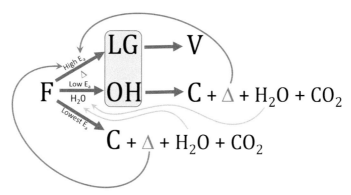

Figure 1.4 Chemical and thermal feedbacks are key features of the thermal degradation of cellulose. The nature and extent of these feedbacks will determine which thermal degradation pathway dominates. Heat released by the char formation process may drive thermal degradation into the high activation energy and endothermic volatilization pathway. Moisture released by the same process may drive thermal degradation into the low activation energy and exothermic char formation pathway.

pathways in cellulose begins to become evident. Heat released from the exothermic formation of char may feed back into the reaction to the high activation energy pathway, leading to the formation of levoglucosan-end cellulose and, subsequently, volatiles. Conversely, the release of water molecules from the dehydration process in the char formation may feed back into the reaction to the water-catalyzed lower activation energy hydrolyzed cellulose and char formation pathway. Thus, the determination of which thermal degradation pathway will dominate the thermal degradation of any particular fuel element depends on the conditions prevailing during the reaction in regard to energy available (a function of reaction temperature) and chemical catalysis.

Under combustion initiation conditions, thermal degradation may commence with char formation. If sufficient heat from this exothermic process is released it may be enough to power the reaction into the volatilization pathway. However, since the volatilization pathway is endothermic, heat will be lost from the system as soon as it begins which may push it back into a lower activation energy charring pathway. If the volatilization pathway is to continue to be maintained, an additional source of heat is required. This additional heat may be supplied by oxidation of the thermal degradation products.

Oxidation of the thermal degradation products (i.e. volatile and char) in the form of flaming and glowing/smoldering combustion introduces additional thermal and chemical feedbacks (Figure 1.5). However, these thermal and chemical elements occur in both solid and gas phase and thus their contributions to the thermal degradation processes are not as straightforward as those considered previously.

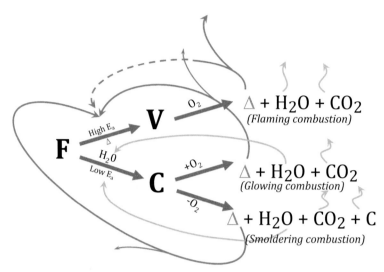

Figure 1.5 Oxidation of the cellulose thermal degradation products creates a new set of chemical and thermal feedbacks that can influence the thermal degradation pathway as well as the propagation of the combustion reactions through new fuel. Oxidation of volatiles (V) in the form of flaming combustion happens in the gas phase and so occurs well above the fuel bed, due to buoyancy, and much of the energy released is transferred outside the immediate combustion zone. In contrast, the solid phase oxidation (C) of glowing and smoldering combustion retains most of the energy in the combustion zone.

The gas phase oxidation of volatiles as flame will occur well above the fuel bed due to the buoyancy of the heated gases released during thermal degradation and, thus, the efficiency with which the heat and combustion products are transferred to the fuel bed will be greatly reduced compared to oxidation of solid phase char, which is not as exposed to such mechanisms. Indeed, this gas phase process is directly affected by the free-air stream (i.e. the wind) in and around the combustion zone and is often particularly turbulent. Similarly, the gas–solid interface reactions of the oxidation of char may also be affected by the fluid dynamics in and around the combustion zone.

However, the effect of the interaction with the free air stream in and around the combustion zone is not just in the impact on the transfer of heat from the gas phase/gas phase and gas phase/solid phase oxidation reactions to the fuel bed but also the active cooling of the fuel bed and combustion zone. The temperature of the free-air stream will be that of the ambient environment, particularly in the initiation phase of combustion, and this dramatically cooler air (compared to that of the combustion zone within the fuel bed) can act to remove heat from the thermally degrading fuel via convective cooling. Thus, the free-air stream acts to disperse the heat and products released in the flaming combustion reactions but also to remove heat from the thermally degrading fuels.

The solid phase/gas phase oxidation of the char is likely to be less affected by the dispersal of heat and convective cooling of the wind because it is within the combustion zone for the most part and, thus, sheltered. However, given the relatively slow rate of these reactions, when the char does oxidize, the heat that is released from these reactions is likely to be at some distance from the leading edge of the combustion zone and thus not play a large role in fires spreading with the wind. At the rear of the fire, however, where the combustion zone is propagating into the wind, the char oxidation process is likely to be more open to the effects of the ambient wind, which may have a significant effect on combustion and, thus, the behavior of the fire on this segment of the perimeter.

The nature of the interaction of the ambient wind field with the combustion zone of a free-spreading fire will depend on the location of the section of the perimeter in question in relation to the direction of the wind. Sections of the fire perimeter that are open to the full influence of the ambient wind, such as the rear of a fire, will have more effective convective cooling of the combustion reactions and thus be more likely to be dominated by the charring and glowing/smoldering pathway. Sections of the perimeter that are somewhat sheltered from the effect of the ambient wind, such as the head of a fire, will have less effective convective cooling, be hotter, and be more likely to be dominated by the volatilization and flaming pathway. The flanks of a fire perimeter will alternate between both conditions as fine scale changes in wind direction alternate fire behavior at the flank between that of heading fire and backing fire.

1.4 Physical Evidence of Environmental Interactions with Combustion Processes

This penultimate section provides some interpreted physical evidence for both the combustion processes discussed in Section 1.2 as well as the interactions of these processes with the environment around them, as discussed in Section 1.3. These are sourced from studies of large-scale free-burning experimental fires as well as small-scale combustion wind tunnel experiments.

1.4.1 Fire Perimeter Propagation

The primary outcome of the environmental interactions with the chemistry of combustion, particularly for a free-spreading fire burning in uniform fuels under consistent conditions, is changes in the pathways of combustion and the resulting behavior of fire around the fire perimeter. Figure 1.6 shows a time-series of aerial

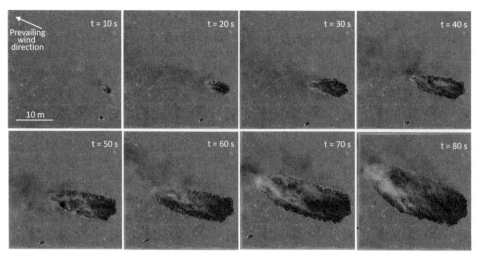

Figure 1.6 Time-series of aerial photographs of the growth of experimental fire Braidwood E14 in grass fuels ignited at a point in a 33 × 33 m plot, commencing 10 seconds after ignition. The growth of the fire in area, perimeter length, and forward rate of spread increases with time as the fire responds to slight changes in the direction of the wind. Conditions for experiment were: grass height 20 cm, grass curing 90%, dead fuel moisture content 7.5% (oven-dry weight), air temperature 301 K (28°C), relative humidity 30%, prevailing mean wind speed at 10 m 17.9 km h^{-1}, maximum interval rate of spread 2.16 km h^{-1}.
A black and white version of this figure will appear in some formats. For the color version, please refer to the plate section.

photographs taken from a remotely-piloted aircraft of an experimental fire burning in continuous grassy fuels after being ignited at a point (Cruz et al. 2015). From the moment of ignition, the fire exhibits nonisotropic spreading behavior determined by the direction of the wind, with the downwind sections of the fire perimeter spreading faster than those burning upwind. Despite the uniformity of the fuel, the direction of the wind imposes a difference in the combustion processes and the formation of flaming combustion. These differences then subsequently lead to a disparity in the heat release and propagation rates around the fire perimeter. The fire quickly develops the typical elliptical shape associated with free-burning fires that steadily grows in size, with slight changes in shape as it responds to minor changes in wind direction.

The rear of the fire, most open to the prevailing ambient wind, exhibits the least energetic flame and fire behavior. The head of the fire, sheltered somewhat from the ambient wind by the presence of the fire, the plume, and the burnt ground upwind of it, exhibits the most energetic flame and fire behavior. This inequality between the upwind and downwind sections of the perimeter is clearly the result of the different heat release and transfer rates, but the cause of these differences is solely the consequence of the competition between the different thermal degradation pathways. The flank sections of the fire perimeter alternate between the behavior of heading and backing fires, depending on minute shifts in the direction of the wind (Cheney 2008).

In the latter stages of this fire, the combustion at the rear of the fire appears for all intents and purposes to have ceased, albeit for a very inconsequential level of flame that is barely, if at all, visible. However, combustion at this section of the perimeter continues and if a significant expedient change in wind direction was to occur, the section would rapidly return to full flaming combustion to become the head.

The changes in combustion dynamics around the fire perimeter appear to be driven primarily by changes in temperature in the combustion zone, driven by the strong temperature dependence of the thermal degradation reactions.

1.4.2 Char Oxidation

The presence of char in a spreading fire is not readily apparent until it undergoes oxidation, either as glowing combustion in the kinetic-controlled regime or smoldering combustion in the diffusion-controlled regime. Even then, the presence of glowing combustion is not easily visible, particularly in the daytime. It is only in comparing the residual combustion behind a spreading fire front during daylight-equivalent illumination with that of semi-darkness that the extent of glowing combustion of char becomes apparent. Figure 1.7 presents images of two free-spreading fires burning through a continuous bed of dry eucalypt forest litter in a large combustion wind tunnel (Sullivan and Matthews 2013; Mulvaney et al. 2016). In these experiments (conducted at different times with slightly different environmental conditions) dead fine fuels comprised of fallen leaves, bark, and twigs (<6 mm diameter) from a local dry eucalypt forest with a fuel load of approximately $1.2 \, \text{kg m}^{-2}$ were burnt in a constant wind of $\simeq 1.5 \, \text{m s}^{-1}$ flowing from right to left. Each fire was ignited from a 1,500-mm-long line at the upwind end of the 4.0 m fuel bed.

Figure 1.7(a), taken approximately 2 minutes 35 seconds after ignition, shows the head fire reaching the end of the fuel bed and the majority of the remainder of the fuel bed converted mostly to ash, evident by the predominance of a paler coloring within the darker color of combustion residue, with some residual burning in denser patches of fuel. Also present in the mostly consumed fuel bed are patches of glowing combustion, evident by the presence of red embers. Figure 1.7(b) shows a very similar scene, albeit for a different fire, but this time with the overhead lights turned off (the lower ambient light levels to which the camera self-adjusted). In this image, much more of the glowing combustion present in Figure 1.7(a) but not apparent is now clearly visible well behind the fire front. Little smoke is visible from these locations of glowing combustion.

(a)

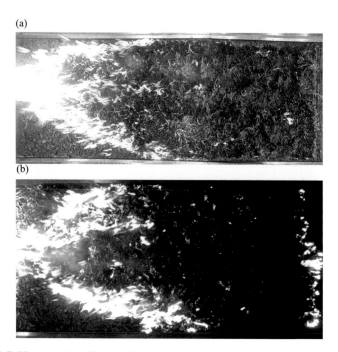

(b)

Figure 1.7 Photographs of two different experimental heading fires burning in dry eucalypt forest litter in the CSIRO Pyrotron combustion wind tunnel with wind blowing right to left. (a) Image taken 155 seconds after ignition. Most of the fuel bed has been consumed with only a small amount of residual flaming and two sources of smoldering combustion visible behind the flame front. (b) Image of another fire burning under similar conditions after the wind tunnel overhead lights have been turned off, revealing a multitude of glowing combustion behind the front.
A black and white version of this figure will appear in some formats. For the color version, please refer to the plate section.

Thus, even when not apparent, there is considerable glowing combustion of char formed in or soon after the passage of the flame front. The slow rate of the solid phase oxidation means that it remains present long after the passage of the flames.

The presence of smoldering combustion is also apparent in these figures as well as those of Figure 1.6, as thin white smoke generated from residual combustion well behind the flame front. Figure 1.8 shows a time-series of aerial photographs taken from a remotely-piloted aircraft of an experimental fire burning in harvested wheat fuels after being ignited along a 33-m-long line on the upwind edge of a 50 m × 50 m plot (Cruz et al. 2020). The fire spreads energetically under a relatively strong wind with flames reaching 3–4 m in height and flames 5–10 m deep. The presence of smoldering combustion in the wake of the passage of the flame zone does not become apparent until more than 37 seconds after ignition. Prior to this there may be glowing combustion but it is not visible in the daylight nor is there any trace of visible emissions as observed in the combustion wind

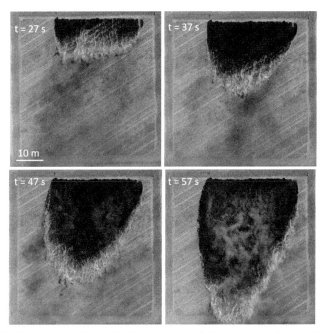

Figure 1.8 Time-series of aerial photographs of the growth of experimental fire Wallinduc WH2 in harvested wheat fuels ignited at a 33-m-long line on the up wind edge of a 50 m × 50 m plot, commencing 27 seconds after ignition. Conditions for experiment were: grass height 20 cm, grass curing 90%, dead fuel moisture content standing fuel 8.7% oven-dry weight, air temperature 303 K (30°C), relative humidity 23%, mean wind speed at 10 m 29.2 $km\ h^{-1}$, maximum interval rate of spread 5.1 $km\ h^{-1}$.

tunnel. Only after time sufficient for the glowing combustion open to the ambient wind upwind of the fire zone to transition from the kinetically-controlled regime to the diffusion-controlled regime does the presence of the char oxidation (visible as the thin white smoke again similar to that in the combustion wind tunnel) become apparent.

1.4.3 Fire Spread Mode: Heading, Backing, Flanking

While it is tempting to assign a particular thermal degradation pathway to a particular mode of fire spread (i.e. heading, backing, or flanking), the stochastic nature of the chemical processes involved does not readily allow this as both thermal degradation pathways – volatilization and charring – may occur simultaneously within the one fuel element (but not the same strand of cellulose). Additionally, as observed in the examples in Figures 1.6–1.8, the subsequent oxidation reactions of one thermal degradation product may obscure the presence of the other. As a result, it may be simpler to use the bulk presence of a particular behavior to ascribe the dominating thermal degradation pathway. In this way, if flaming is dominant, then combustion may be described as being dominated by the volatilization pathway or, if smoldering or glowing combustion is dominant, then charring may dominate combustion. However, such a description does not convey the true nature of the combustion.

Figures 1.9–1.11 illustrate the distinct natures of the combustion of heading, flanking, and backing fires, respectively, burning in dry eucalypt litter under constant conditions of a 1.5 m s^{-1} wind in a combustion wind tunnel. Figure 1.9 shows a time series of a heading fire propagating with the wind at a speed of about 80 m h^{-1}. Figure 1.10 shows a time series of a flanking fire propagating parallel to the wind at a speed of about 6 m h^{-1}. Figure 1.11 shows a time series of a backing fire propagating against the wind at a speed of about 5 m h^{-1}. The nature of the changes in the combustion chemistry resulting from the nonisotropic forcing of the thermal degradation kinetics is clearly apparent, with the heading fire producing much more energetic flames.

The flanking and backing fires have flames of similar dimension and spread rate, with the flames of the flank fire being less uniform and spreading slightly faster than those of the backing fire. The oxidation of charred fuel formed in the

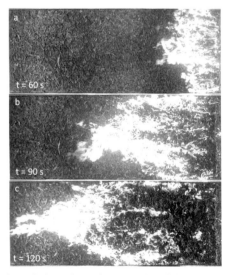

Figure 1.9 Time-series of plan-view photographs of the growth of an experimental heading fire burning in dry eucalypt litter in a 1.5 m s^{-1} wind and moisture content of 7–8% oven dry weight. The fire was ignited from a 1.5-meter line on the upwind edge of a 4-meter fuel bed. Average rate of spread was 80 m h^{-1}. Air flow is from right to left: (a) 60 seconds after ignition, (b) 90 seconds after ignition, (c) 120 seconds after ignition.

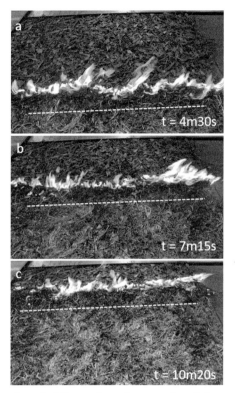

Figure 1.10 Time-series of oblique view photographs of the growth of an experimental flank fire burning in dry eucalypt litter in a 1.5 m s^{-1} wind and moisture content of 7–8% oven dry weight. The fire was ignited from a 1.5 meter-line on the parallel edge of a 1.5-meter fuel bed and had a mean spread rate of 6 m h^{-1} from bottom to top. Air flow is from left to right: (a) 4 minutes 30 seconds after ignition, (b) 7 minutes 15 seconds after ignition, (c) 10 minutes 20 seconds after ignition. The white dashed line roughly indicates the divide between the char zone undergoing oxidation and the ash zone that has completed oxidation.

A black and white version of this figure will appear in some formats. For the color version, please refer to the plate section.

flame zone is recognizable in the progression of the zone of charred fuel immediately behind the flame zone immediately followed by a zone of white ash, delineated in Figure 1.10 by a dashed white line. The presence of such a zone of charred fuel is not so apparent in the images of the backing fire, possibly due to the obscuring of it by the flames leaning over the burnt fuel and also by the much slower rate of progression of the flame front, giving the false impression that fuels are converted directly to char.

While distinct differences in the behavior and spread can be seen in each fire spread mode, the impact on the residual ash is less obvious. Figure 1.12 shows images of the fuel bed taken after the completion of each type of fire. No

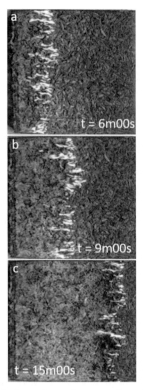

Figure 1.11 Time-series of plan-view photographs of the growth of an experimental backing fire burning in dry eucalypt litter in a 1.5 m s^{-1} wind and moisture content of 6–7% oven dry weight. The fire was ignited from a 1.5-meter line on the downwind edge of a 2-meter fuel bed and had a mean spread rate of 5 m h^{-1}. Air flow is from right to left, spread is left to right: (a) 6 minutes after ignition, (b) 9 minutes after ignition, (c) 15 minutes after ignition.

discernable difference can be easily detected between the fire spread modes illustrated here. This is primarily because, after combustion is completed, either flaming or glowing/smoldering oxidation, all fuels are generally equally consumed regardless of spread mode. This may not be the case where conditions are more marginal or the combustion process is incomplete. The fact that a char zone is visible in the flank fire spread, between the flame zone and ash zone, suggests that if the residual glowing/smoldering combustion was interrupted in some way, then the char zone would remain.

1.4.4 Emissions

It has been generally accepted (Andreae and Merlet 2001; Andreae 2019) that smoldering combustion is associated with emissions of CO and flaming

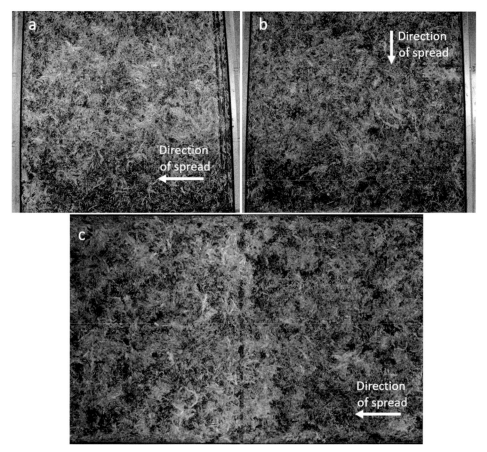

Figure 1.12 Plan views of the dry eucalypt litter ash beds after burning in a combustion wind tunnel by different fire spread modes: (a) flanking fire, (b) backing fire, (c) heading fire.

combustion is associated with emissions of CO_2. And, since smoldering combustion is generally associated with low intensity fires such as backing fires and flaming combustion is generally associated with heading fires, then backing fires should, by inference, produce more CO than heading fires, and heading fires should produce more CO_2 than backing fires. However, the lack of distinct difference in ash observed in Figure 1.12 is also mirrored by analysis of emissions factors for carbon species (i.e. CO_2, CO, and CH_4) by fire spread mode, as shown in Figure 1.13 and reported by Surawski et al. (2015).

It can be seen that heading fires produce both more CO_2 and more CO than flanking or backing fires. Backing and flanking fires produce roughly similar amounts of these emissions. In contrast, heading fires produce significantly less carbon residue than either flanking or backing fires, a result of the consumption of

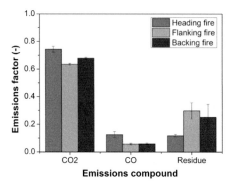

Figure 1.13 Carbon emissions factors, CO_2 and CO, and burnt carbon residue (g kg^{-1} fuel consumed) for different fire spread modes. Source: Surawski et al. (2015)

more biomass more generally in oxidation reactions as well as a greater tendency to continue combustion through to completion. While the flashy flames of a head fire suggest all combustion is driven by the production of volatiles, it is clear that the increased temperatures associated with such spread also result in increased reaction rates for production of char. The rapid propagation of the flame zone of most head fires leaves most of the slower oxidation of the char to happen well behind the flame zone and thus exposed to the cooler ambient air flowing over the previously burnt ground, as seen in Figures 1.8 and 1.9. This then results in the char oxidizing in the kinetically controlled regime, producing more CO.

In contrast, the formation of both the volatile and char in flanking and backing fires is much reduced due to the relatively cooler combustion environment and the exposure to ambient-temperature wind resulting in lower temperatures in the combustion zone and thus lower reaction rates, leaving a greater proportion of the original fuel as not fully combusted in the form of pyrogenic carbon (Surawski et al. 2020). The subsequent oxidation as both flaming and smoldering or glowing combustion also tends to be more confined, less exposed to cooling ambient conditions, and thus more complete.

1.5 Concluding Remarks

We have seen that both the chemistry of biomass fuel and the nature of its combustion through thermal degradation and oxidation reactions is quite complex. The competitive thermodynamics of the thermal degradation reactions as a direct consequence of the nucleophilic competition between the volatilization pathway and the charring pathway as well as differences in the activation energies and enthalpies of these pathways, the high level of sensitivity of the rates of these reactions to reactant temperature, and differences in the magnitudes and locations of heat released during oxidation explain much of the apparent capriciousness of biomass combustion.

The interaction of this complex set of competing and interacting reactions with the often-changing environment around a fire, particularly that of the ambient wind, results in a highly dynamic combustion situation peculiar to fires burning in the open.

It is the interaction of these factors than help explain many of the observed behaviors of free-burning wildland fires such as the shape of a fire's perimeter, the change in behavior around that perimeter with respect to the direction of the wind, the influence the wind has on the behavior and spread of a fire, and the importance of the different combustion pathways in sustaining spread, particularly under marginal conditions.

It is also the interaction of these factors that help explain our initial question as to why blowing on the embers of a fire when it is being lit helps. As we have seen, the actual initiating reactions are anaerobic and so blowing does not critically supply oxygen. Blowing acts to cool the incipient reactions sufficiently to drive the reaction into the charring pathway. This pathway, being exothermic, generates more heat (and, yes, some oxidation of the char as glowing combustion) which, when the blowing stops, is hopefully enough to push the reaction process into the endothermic volatilization pathway and subsequent flaming combustion.

If the blowing is too vigorous it can cool the degrading fuel too much and stop the thermal degradation reactions. If the blowing is not vigorous enough or stops too early, the transition of the reactions to volatilization will remove too much heat from the system and act to stop that thermal degradation pathway.

If we return to our initial diagram of combustion and sustained spread (Figure 1.1, we can see that it is not sophisticated enough to capture this fundamental aspect of biomass combustion, particularly for understanding the behavior and spread of free-burning fires in wildland fuels. Figure 1.14 is an attempt to revise this diagram to include these aspects but still be relatively simple.

In this revised conceptual model, fuel still undergoes drying from an ignition or pilot source. Under continued application of this heat, the fuel thermally degrades, either into hydrolyzed cellulose under the exothermic, lower activation energy reaction pathway, or into levoglucosan-end via the endothermic, higher activation energy pathway. If sufficient heat is created in the hydrolyzed cellulose pathway, the reaction may be pushed toward the endothermic pathway. Continued heating results in either char from the hydrolyzed cellulose or volatilization of the levoglucosan-end cellulose. Again, if enough heat is generated from the charring pathway, the thermokinetics may be driven into the levoglucosan-end and volatilization pathway. Conversely, if too much heat is lost from the volatilization

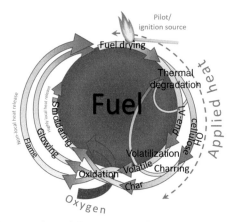

Figure 1.14 A revised version of the conceptual model presented in Figure 1.1 with the distance from the center of the circle indicating the magnitude of enthalpy. In this conceptual model, the fundamental competitive thermokinetics between the formation of levoglucosan-end cellulose (LV-end) and hydrolyzed cellulose (OH cellulose) during thermal degradation drives volatilization and the formation of volatile and charring and the formation of char, respectively. Heat formed in the production of hydrolyzed cellulose or char may be sufficient to drive the reaction toward the endothermic volatilization pathway. These thermal degradation products may then undergo oxidation in the form of flame in the case of volatile (which as a gas–gas reaction is open to turbulent mixing) and glowing or smoldering combustion in the case of char. The heat released by glowing or smoldering is highly localized within the fuel bed, while that released by flame may be some distance from the fuel bed.

A black and white version of this figure will appear in some formats. For the color version, please refer to the plate section.

pathway, the thermokinetics may be driven back into the hydrolyzed cellulose pathway, or thermal degradation may cease completely.

The thermal degradation products may then oxidize in the presence of oxygen, liberating much more heat into the system. In the case of volatiles, this is a gas–gas reaction and thus open to turbulent mixing, which may promote or inhibit oxidation. This heat is generally released well away from the fuel bed where the volatile was formed and increases the amount of adjacent fuel exposed to heating and thus the rate of spread of the fire. In the case of char, oxidation follows either the kinetic-controlled regime in which the char oxidizes in smoldering combustion or the diffusion-controlled regime in which the char oxidizes in glowing combustion. The amount of heat released in the former is much less than the latter and little of it may result in heating of adjacent fuels. In either case, the rate of propagation of a fire edge is much slower than that of flaming combustion.

Understanding the factors that determine the amount of heat released and rate at which it is transferred to adjacent fuel in order to predict the speed at which a fire may propagate is the primary objective of wildland fire behavior science and necessarily involves a broad range of disciplines, including radiant and convective heat transfer, fluid dynamics, and meteorology, many of which are covered in other chapters of this textbook.

References

Anderson, HE, Rothermel, RC (1965) Influence of moisture and wind upon the characteristics of free-burning fires. *Symposium (International) on Combustion* **10**(1), 1009–1019.

Andreae, MO (2019) Emission of trace gases and aerosols from biomass burning: An updated assessment. *Atmospheric Chemistry and Physics* **19**(13), 8523–8546.

Andreae, MO, Merlet, P (2001) Emission of trace gases and aerosols from biomass burning. *Global Biogeochemical Cycles* **15**(4), 955–966.

Antal, MJ, Várhegyi, G (1995) Cellulose pyrolysis kinetics: The current state of knowledge. *Industrial & Engineering Chemistry Research* **34**(3), 703–717.

Antal, MJ, Várhegyi, G, Jakab, E (1998) Cellulose pyrolysis kinetics: Revisited. *Industrial & Engineering Chemistry Research* **37**(4), 1267–1275.

Atreya, A (1998) Ignition of fires. *Philosophical Transactions of the Royal Society A: Mathematical, Physical and Engineering Sciences*, **356**(1748), 2787–2813.

Babrauskas, V (2003) *Ignition Handbook*. Issaquah, WA: Fire Science Publishers.

Bachelard, G (1938) *The Psychoanalysis of Fire*. London: Quartet Books. (orig. *La Psychanalyse du Feu*, trans. Alan C.M. Ross, 1964).

Ball, R, McIntosh, AC, Brindley, J (1999a) The role of char-forming processes in the thermal decomposition of cellulose. *Physical Chemistry Chemical Physics* **1**(21), 5035–5043.

Ball, R, McIntosh, AC, Brindley, J (1999b) Thermokinetic models for simultaneous reactions: A comparative study. *Combustion Theory and Modelling* **3**(3), 447–468.

Ball, R, McIntosh, AC, Brindley, J (2004) Feedback processes in cellulose thermal decomposition: Implications for fire-retarding strategies and treatments. *Combustion Theory and Modelling* **8**(2), 281–291.

Beall, FC, Eickner, HW (1970) Thermal Degradation of Wood Components: A Review of the Literature. Research Paper FPL 130. USDA Forest Service, Madison, Wisconsin.

Bilger, RW (1989) Turbulent diffusion flames. *Annual Review of Fluid Mechanics* **21**(1), 101–135.

Boonmee, N, Quintiere, JG (2005) Glowing ignition of wood: The onset of surface combustion. *Proceedings of the Combustion Institute* **30**(2), 2303–2310.

Bowman, CT (1975) Non-equilibrium radical concentrations in shock-initiated methane oxidation. *Symposium (International) on Combustion* **15**(1), 869–882.

Branca, C, di Blasi, C (2004) Parallel- and series-reaction mechanisms of wood and char combustion. *Thermal Science* **8**(2), 51–63.

Broido, A (1976) Kinetics of solid-phase cellulose pyrolysis. In: Shafizadeh, F, Sarkanen, KV, Tillman, DA, eds. *Thermal Uses and Properties of Carbohydrates and Lignins*. New York: Academic Publishing, pp. 19–36.

Broido, A, Javierson, AC, Ouano, AC, Barrall II, EM (1973) Molecular weight decrease in the early pyrolysis of crystalline and amorphous cellulose. *Journal of Applied Polymer Science* **17**(12), 3627–3635.

Broido, A, Weinstein, M (1970) Thermogravimetric analysis of ammonia-swelled cellulose. *Combustion Science and Technology* **1**(4), 279–285.

Cheney, NP (2008) Can forestry manage bushfires in the future? *Australian Forestry* **71**(1), 1–2.

Cheney, P, Sullivan, A (2008) *Grassfires: Fuel, Weather and Fire Behaviour*, 2nd ed. Collingwood, Australia: CSIRO Publishing.

Coblentz, WW (1905) Infra-red absorption spectra, II: Liquids and solids. *The Physical Review* **20**(6), 337–363.

Cruz, MG, Gould, JS, Kidnie, S, Bessell, R, Nichols, D, Slijepcevic, A (2015) Effects of curing on grassfires: II. Effect of grass senescence on the rate of fire spread. *International Journal of Wildland Fire* **24**(6), 838–848.

Cruz, MG, Hurley, RJ, Bessell, R, Sullivan, AL (2020) Fire behaviour in wheat crops: Effect of fuel structure on rate of fire spread. *International Journal of Wildland Fire* **29**(3), 258–271.

Deguchi, S, Tsujii, K, Horikoshi, K (2006) Cooking cellulose in hot and compressed water. *Chemical Communications* **11**(31), 3293–3295.

Demirbaş, A (2004) Combustion characteristics of different biomass fuels. *Progress in Energy and Combustion Science* **30**(2), 219–230.

Di Blasi, C (1993) Modeling and simulation of combustion processes of charring and non-charring solid fuels. *Progress in Energy and Combustion Science* **19**(1), 71–104.

Di Blasi, C (1998) Comparison of semi-global mechanisms for primary pyrolysis of lignocellulosic fuels. *Journal of Analytical and Applied Pyrolysis* **47**(1), 43–64.

Diebold, JP (1994) A unified, global model for the pyrolysis of cellulose. *Biomass and Bioenergy* **7**(1–6), 75–85.

Eghlimi, A, Lu, L, Sahajwalla, V, Harris, D (1999) Computational modelling of char combustion particles based on the structure of char particles. In: *Second International Conference of CFD in the Minerals and Process Industries*; December 6–8, 1999. CSIRO, Melbourne, Australia.

Gaydon, AG, Wolfhard, HG (1960) *Flames: Their Structure, Radiation and Temperature*, 2nd ed. London: Chapman and Hall Ltd.

Gisborne, HT (1948) Fundamentals of fire behavior. *Fire Control Notes* **9**(1), 13–24.

Gordobil, O, Moriana, R, Zhang, L, Labidi, J, Sevastyanova, O (2016) Assessment of technical lignins for uses in biofuels and biomaterials: Structure-related properties, proximate analysis and chemical modification. *Industrial Crops and Products* **83**(May), 155–165.

Gould, JS, McCaw, WL, Cheney, NP (2011) Quantifying fine fuel dynamics and structure in dry eucalypt forest (Eucalyptus marginata) in Western Australia for fire management. *Forest Ecology and Management* **262**(3), 531–546.

Greenberg, JP, Friedli, H, Guenther, AB, Hanson, D, Harley, P, Karl, T (2006) Volatile organic emissions from the distillation and pyrolysis of vegetation. *Atmospheric Chemistry and Physics* **6**(1), 81–91.

Grootemaat, S, Wright, IJ, van Bodegom, PM, Cornelissen, JHC (2019) Scaling up flammability from individual leaves to fuel beds. *Oikos* **126**(10), 1428–1438.

Gullett, BK, Touati, A (2003) PCDD/F emissions from forest fire simulations. *Atmospheric Environment* **37**(6), 803–813.

Harris, P (1999) On charcoal. *Interdisciplinary Science Reviews* **24**(4), 301–306.

Howard, D, Macsween, K, Edwards, GC, Desservettaz, M, Guérette, E-A, Paton-Walsh, C, Surawski, NC, Sullivan, AL, Weston, C, Volkova, L, Powell, J, Keywood, MD, Reisen, F, (Mick) Meyer, CP (2019) Investigation of mercury emissions from burning of Australian eucalypt forest surface fuels using a combustion wind tunnel and field observations. *Atmospheric Environment* **202**(Apr.), 17–27.

Hurst, DF, Griffith, DWT, Cook, GD (1994) Trace gas emissions from biomass burning in tropical Australian savannas. *Journal of Geophysical Research* **99**(D8), 16441–16456.

Jane, FW (1956) *The Structure of Wood*. New York: The Macmillan Company.

Lemieux, PM, Lutes, CC, Santoianni, DA (2004) Emissions of organic air toxics from open burning: A comprehensive review. *Progress in Energy and Combustion Science* **30**(1), 1–32.

Luke, RH, McArthur, AG (1978) *Bushfires in Australia*. Canberra: Australian Government Publishing Service.

McArthur, AG (1967) *Fire Behaviour in Eucalypt Forests*. Forestry and Timber Bureau Leaflet 107. Canberra: Commonwealth Department of National Development.

Milosavljevic, I, Oja, V, Suuberg, EM (1996) Thermal effects in cellulose pyrolysis: Relationship to char formation process. *Industrial & Engineering Chemistry Research* **35**(3), 653–662.

Mok, WSL, Antal, MJ (1983) Effects of pressure on biomass pyrolysis. II. Heats of reaction of cellulose pyrolysis. *Thermochimica Acta* **68**(2–3), 165–186.

Moore, WJ (1963) *Physical Chemistry*, 4th ed. London: Longmans Green and Co Ltd.

Morrison, RT, Boyd, RN (1983) *Organic Chemistry*, 4th ed. Boston, MA: Allyn and Bacon, Inc.

Mulvaney, JJ, Sullivan, AL, Cary, GJ, Bishop, GR (2016) Repeatability of free-burning fire experiments using heterogeneous forest fuel beds in a combustion wind tunnel. *International Journal of Wildland Fire* **25**(4), 445–455.

Ohlemiller, TJ (1985) Modeling of smoldering combustion propagation. *Progress in Energy and Combustion Science* **11**(4), 277–310.

O'Sullivan, AC (1997) Cellulose: The structure slowly unravels. *Cellulose* **4**(3), 173–207.

Parker, WJ, LeVan, SL (1989) Kinetic properties of the components of douglas-fir and the heat of combustion of their volatile pyrolysis products. *Wood and Fiber Science* **21**(3), 289–305.

Philibert, J (2006) Some thoughts and/or questions about activation energy and pre-exponential factor. *Defect and Diffusion Forum* **249**, 61–72.

Pyne, SJ, Andrews, PL, Laven, RD (1996) *Introduction to Wildland Fire*, 2nd ed. New York: John Wiley and Sons.

Rein, G (2013) Smouldering fires and natural fuels. In: Belcher, CM, ed. *Fire Phenomena and the Earth System*. New York: John Wiley & Sons, pp. 15–33.

Rein, G (2016) Smoldering Combustion. In: Hurley, MJ, Gottuk, DT, Hall Jr., JR, Harada, K, Kuligowski, ED, Puchovsky, M, Torero, JL, Watts Jr., JM, Wieczorek, CJ, eds. *SFPE Handbook of Fire Protection Engineering*. New York: Springer, pp. 581–603.

Reisen, F, Brown, S, Cheng, M (2006) Air toxics in bushfire smoke: Firefighters exposure during prescribed burns. *Forest Ecology and Management* **234**(Supp 1), S144–S144.

Rothermel, RC (1972) *A Mathematical Model for Predicting Fire Spread in Wildland Fuels*. Research Paper INT-115. USDA Forest Service, Intermountain Forest and Range Experimental Station, Odgen UT.

Shafizadeh, F (1982) Introduction to pyrolysis of biomass. *Journal of Analytical and Applied Pyrolysis* **3**(4), 283–305.

Sullivan, AL (2017a) Inside the inferno: Fundamental processes of wildland fire behaviour. Part 1: Combustion chemistry and energy release. *Current Forestry Reports* **3**, 132–149.

Sullivan, AL (2017b) Inside the inferno: Fundamental processes of wildland fire behaviour. Part 2: Heat transfer and interactions. *Current Forestry Reports* **3**, 150–171.

Sullivan, AL, Ball, R (2012) Thermal decomposition and combustion chemistry of cellulosic biomass. *Atmospheric Environment*, **47**(Feb.), 133–141.

Sullivan, AL, Matthews, S (2013) Determining landscape fine fuel moisture content of the Kilmore East "Black Saturday" wildfire using spatially-extended point-based models. *Environmental Modelling and Software* **40**, 98–108.

Surawski, NC, Macdonald, LM, Baldock, JA, Sullivan, AL, Roxburgh, SH, Polglase, PJ (2020) Exploring how fire spread mode shapes the composition of pyrogenic carbon from burning forest litter fuels in a combustion wind tunnel. *Science of the Total Environment*, **698**(Jan.), 134306.

Surawski, NC, Sullivan, AL, Meyer, CP, Roxburgh, SH, Polglase, PJ (2015) Greenhouse gas emissions from laboratory-scale fires in wildland fuels depend on fire spread mode and phase of combustion. *Atmospheric Chemistry and Physics* **15**(9), 5259–5273.

Surawski, NC, Sullivan, AL, Roxburgh, SH, Polglase, PJ (2016) Estimates of greenhouse gas and black carbon emissions from a major Australian wildfire with high spatiotemporal resolution. *Journal of Geophysical Research: Atmospheres* **121**(16), 9892–9907.

Varner, JM, Kane, JM, Kreye, JK, Engber, E (2015) The flammability of forest and woodland litter: A synthesis. *Current Forestry Reports* **1**(2), 91–99.

Vervisch, L, Poinsot, T (1998) Direct numerical simulation of non-premixed turbulent flames. *Annual Review of Fluid Mechanics*, **30**, 655–691.

Vines, RG (1981) Physics and chemistry of rural fires. In: Gill, AM, Groves, RH, Noble, IR, eds. *Fire and the Australian Biota*. Canberra: Australian Academy of Science, pp. 129–150.

Weinstein, M, Broido, A (1970) Pyrolysis–crystallinity relationships in cellulose. *Combustion Science and Technology* **1**(4), 287–292.

Williams, FA (1977) Mechanisms of fire spread. *Symposium (International) on Combustion* **16**(1), 1281–1294.

Williams, FA (1982) Urban and wildland fire phenomenology. *Progress in Energy Combustion Science* **8**(4), 317–354.

Wodley, FA (1971) Pyrolysis products of untreated and flame retardant-treated a-cellulose and levoglucosan. *Journal of Applied Polymer Science* **15**(4), 835–851.

Yaman, S (2004) Pyrolysis of biomass to produce fuels and chemical feedstocks. *Energy Conversion and Management* **45**(5), 651–671.

2

The Structure of Line Fires at Flame Scale

SALMAN VERMA, MOHAMED AHMED, AND ARNAUD TROUVÉ

2.1 Introduction

The dynamics of wildland fires involve multi-physics phenomena occurring at multiple scales. Different length scales are believed to play a role in fire behavior: the vegetation scales, denoted $L_{vegetation}$, that characterize the biomass fuel; the flame scales represented by a characteristic flame height and width, L_{flame} and W_{flame}, as well as the length of the fire line, $L_{fireline}$, that characterize the combustion and heat transfer processes; the geographical scales, $L_{topography}$ and L_{land_cover}, that characterize the terrain topography ($L_{topography}$ is the characteristic scale for change in elevation of the terrain) and land cover (L_{land_cover} is the characteristic scale for change in vegetation type and/or surface fuel load); and the meteorological scales represented by the depth of the atmospheric boundary layer, L_{ABL}, that characterize atmospheric conditions. In addition, the fire plume has scales that can be represented by a characteristic height and width, L_{plume} and W_{plume}; the plume scales take a large range of values as they grow from flame scales to geographical scales and then to meteorological scales. In wildland fire problems, $L_{vegetation}$ is on the order of a few millimeters or centimeters; L_{flame} and W_{flame} are on the order of a few meters; $L_{fireline}$, $L_{topography}$, and L_{land_cover} are typically on the order of a few tens or hundreds of meters; and L_{ABL} is on the order of kilometers.

Computational Fluid Dynamics (CFD) models have the potential to provide detailed information on the interactions between physical phenomena occurring at all these different scales. However, because of computational cost, the domain of application of CFD models is typically limited to a particular range of scales. Thus, current CFD-based wildland fire models are scale-specific and belong to one of the following three classes (see Figure 2.1): combustion solvers aimed at describing

Results presented in this chapter were produced in the past few years through projects financially supported by the U.S. Forest Service, Rocky Mountain Research Station, and by FM Global. These projects were also supported by supercomputing resources made available by the University of Maryland (http://hpcc.umd.edu) and by the National Science Foundation (XSEDE Program, Grant #TG-CTS140046).

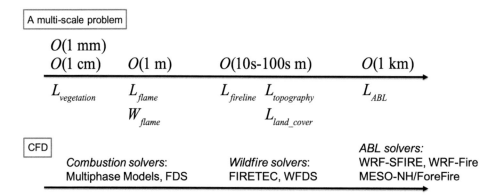

Figure 2.1 The different classes of CFD models used for wildland fire spread simulations: combustion solvers resolve dynamics at the vegetation and flame scales; wildfire solvers resolve dynamics at the fireline and geographical scales; atmospheric boundary layers (ABL) solvers resolve dynamics at the meteorological scales.

the coupling between pyrolysis, combustion, radiation, and flow occurring at the vegetation and flame scales; wildfire solvers aimed at describing the coupling between combustion and flow occurring at fire line scales and/or geographical scales; and atmospheric boundary layer solvers aimed at describing the coupling between combustion and flow occurring at meteorological scales.

Examples of combustion solvers that have been developed for wildland fire dynamics applications include a group of models known as multiphase models (Porterie et al. 2000; Morvan and Dupuy 2001, 2004). These solvers use a computational grid resolution of order 1–10 cm and provide a fine-grained treatment of the pyrolysis, combustion, and heat transfer processes that are responsible for flame spread through a first principles-based model. Simulations with these solvers are typically performed in small domains (a few tens of meters in two-dimensional simulations or a few meters in three-dimensional simulations). Other examples of combustion solvers include FDS[1] and FireFOAM[2]; these solvers are well-established fire models that are primarily used for building fire applications; they have also been adapted for wildland fire applications.

Examples of wildfire solvers include FIRETEC (Linn and Cunningham 2005; Canfield et al. 2014) and WFDS (Mell et al. 2007) (WFDS is based on FDS). These solvers use a computational grid resolution of order 1 m and provide a coarse-grained treatment of unresolved vegetation-scale and flame-scale processes through a simplified (but physics-based) combustion model. Simulations with

[1] Fire Dynamics Simulator (FDS), developed by the National Institute of Standards and Technology, n.d. https://pages.nist.gov/fds-smv/ (last accessed September 29, 2019).
[2] FireFOAM, developed by FM Global, n.d. https://github.com/fireFoam-dev (last accessed September 29, 2019).

these solvers are typically performed in intermediate-size field-scale domains (e.g. 1 km in size).

Examples of atmospheric boundary layer (ABL) solvers that have been developed for wildland fire dynamics applications include WRF-SFIRE and WRF-Fire (Clark et al. 2004; Mandel et al. 2011; Coen et al. 2013; Kochanski et al. 2013), as well as MESO-NH/ForeFire (Filippi et al. 2009, 2013). These solvers use a computational grid resolution of order 10–100 m and provide a macroscopic-level treatment of unresolved vegetation-scale, flame-scale, fireline-scale, and small-geographical-scale processes through a parametrized semi-empirical rate-of-spread fire model. Simulations with ABL solvers are typically performed in arbitrary-size field-scale domains (from a few kilometers to several tens of kilometers and beyond). A strength of ABL solvers is that they are integrated with research-level or operational-level numerical weather prediction capabilities (i.e. WRF and MESO-NH) and therefore incorporate detailed descriptions of the fuel maps, topographic maps, and weather conditions.

In the present chapter, we focus on the viewpoint adopted in combustion solvers and use high-resolution computational modeling to bring basic information on the coupling between combustion, radiation, and turbulent flow processes occurring at flame scale. This viewpoint is typically adopted in standalone theoretical studies or in joint computational–experimental studies of controlled laboratory-scale flames. In addition, we limit the scope of the discussion to a simplified configuration in which the flame is fueled by methane gas supplied at a prescribed flow rate. In this configuration, the heat release rate of the flame is controlled and the flame is stationary (i.e. non-spreading). It is worth emphasizing that a number of important features of wildland fires are left out of the present gas-fueled configuration; for instance: the presence of a complex and discontinuous system of vegetation fuel in the wildland in the form of discrete and disconnected particles of different shapes and sizes; the uncontrolled production of flammable vapors (i.e. the fuel) due to pyrolysis reactions occurring inside the vegetation fuel and driven by the gas-to-solid (convective and radiative) heat transfer; the typical cycle of ignition, pyrolysis and end-of-pyrolysis (due to a reduction in the gas-to-solid heat transfer) or burn-out (due to fuel depletion); and the displacement (i.e. the spreading) of the flame region. The study of gas-fueled configurations may be viewed as an intermediate step on the road to provide a basic understanding of the dynamics of wildland fires.

Furthermore, we focus in the present chapter on the line fire configuration. Line fires have a simplified structure (they are statistically two-dimensional) and are representative of the flame geometry found in wildland fire applications in which the fire may be viewed as a thin front that separates unburned and burned vegetation: the flame height and flame width, L_{flame} and W_{flame}, are on the order of a few meters, while the length of the fireline, $L_{fireline}$, is on the order of a few tens or hundreds of meters (or more); we therefore have: $L_{fireline} \gg L_{flame}$ and $L_{fireline} \gg W_{flame}$. Thus, the line fire configuration may be viewed as a canonical configuration in wildland fire research (note that this is perhaps the only canonical configuration currently accepted in wildland fire research). In the following, we consider two variations of the line fire configuration: a first configuration in which the gas-fueled flame develops along a horizontal flat surface under controlled cross-flow conditions that represent wind; and a second configuration in which the gas-fueled flame develops along an inclined flat surface that represents sloped terrain. In the configuration with wind, the direction of the wind is taken as perpendicular to the line flame; in the configuration with slope, the direction of the slope is taken as perpendicular to the line flame.

2.2 The Dynamics of Line Fires

One of the most striking features of flames propagating over a vegetation bed in the presence of a cross-flow or along sloped terrain is the existence of two limiting flame regimes (Pagni and Peterson 1973; Morvan et al. 1998; Morvan and Dupuy 2004; Morvan 2011): the plume-dominated regime in which the flame is mostly detached from the bed, has a vertically tilted geometry and air entrainment is two-sided; and the wind-driven or slope-driven regime in which the flame is attached to the bed, has a boundary layer geometry and air entrainment is one-sided (see Figure 2.2). These two regimes correspond to different flame spread mechanisms

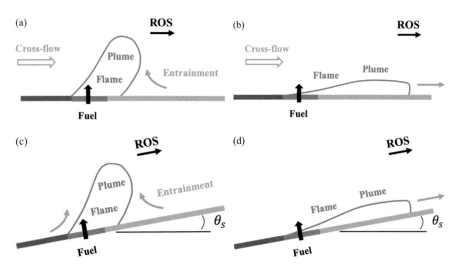

Figure 2.2 The different flame regimes found in wildland fire spread: (a) and (c) the detached flame regime; and (b) and (d) the attached flame regime. Cases (a) and (b) correspond to a configuration with wind; cases (c) and (d) correspond to a configuration with sloped terrain.

(and in particular different relative weights for radiative and convective heat transfer) and to different values of the rate of spread (ROS): in the plume-dominated regime, the surfaces that are downwind/upslope of the flame are exposed to an inflow of ambient air, that is, to convective cooling, and flames spreading in the downwind/upslope direction will therefore feature low-to-moderate values of ROS; in contrast, in the wind-driven or slope-driven regime, the surfaces that are downwind/upslope of the flame are exposed to an outflow of combustion gases, that is, to convective heating, and flames spreading in the downwind/upslope direction will therefore feature moderate-to-high values of ROS. Recent work has shed some new light on the importance of convective heating in controlling fire spread under conditions aided by wind or slope (Finney et al. 2013, 2015). Note also that the possible transition to the limiting wind-driven or slope-driven regime is believed to be a key factor, and possibly the main factor, in a special category of fire behavior known as "eruptive fires" (Viegas 2005; Dold and Zinoviev 2009): eruptive fires are characterized by large values of flame acceleration, high (and time-dependent) values of ROS, and flow and flame attachment to the ground surface (Dold and Zinoviev 2009). On a separate but related subject, variations in the flame structure and in downwind surface conditions similar to those observed in the case of vegetation fires (albeit without spread) have also been observed in the case of liquid pool fires (Lam and Weckman 2015a, 2015b; Hu 2017) and low-momentum gaseous flames (Tang et al. 2017) exposed to a cross-flow.

In the first configuration with wind (and without slope), the literature suggests that the flame structure is controlled by a non-dimensional number known as Byram's convection number and noted N_C (Albini 1981; Raupach 1990; Nelson 1993; Morvan and Frangieh 2008; Nelson et al. 2012):

$$N_C = 2\frac{(1 - \chi_{rad})\dot{Q}'g}{(\rho_a c_{p,a} T_a)u_w^3} \tag{2.1}$$

where $\dot{Q}'$ is the fire intensity (i.e. the rate of heat release due to combustion per unit length of the fireline), χ_{rad} the radiant fraction of the flame (i.e. the fraction of the energy released by combustion that is lost to the surroundings due to thermal radiation transport), g the magnitude of the gravity acceleration, ρ_a, $c_{p,a}$, and T_a the mass density, heat capacity (at constant pressure), and temperature of ambient air, respectively, and u_w the magnitude of the cross-wind velocity. The importance of N_C can be readily understood if one considers that the characteristic buoyant velocity of a line fire without wind is $w_B \sim \left(\dot{Q}'g/(\rho_a c_{p,a} T_a)\right)^{(1/3)}$ (Yuan and Cox 1996); thus N_C gives an estimate of the relative strength of buoyant forces that drive the fire in the vertical direction, as measured by

w_B, and inertial forces that drive the fire in the direction tangent to the (horizontal) ground surface, as measured by u_w, $N_C \sim (w_B/u_w)^3$. High values of N_C correspond to a plume-dominated fire; low values of N_C correspond to a wind-driven fire. The literature suggests that the transition between the two flame regimes occurs for values of N_C between 2 and 10 (Morvan and Frangieh 2008; Nelson et al. 2012).

Note that the literature does not always recognize N_C as the main parameter to characterize the structure of line fires with wind and that alternative parameters, usually cast in the form of a Froude or a Richardson number, have also been proposed. For instance, the studies in Putnam (1965), Morvan and Dupuy (2004), and Nmira et al. (2010) use a Froude number as their main scaling parameter, $Fr = u_w^2/(gL_{f,0})$, where $L_{f,0}$ is the vertical flame height of the line fire without wind. In many cases, these alternative scaling parameters are directly related to Byram's convection number: for instance, if one considers that the flame height of a line fire without wind (or slope) scales like $L_{f,0} \sim \left(\dot{Q}'/(\rho_a c_{p,a} T_a \sqrt{g})\right)^{(2/3)}$ (Yuan and Cox 1996), one finds that $Fr \sim N_C^{(-2/3)}$.

Furthermore, it is worth noting that a few recent studies have questioned the value of Byram's convection number as a controlling factor of wind-aided fire spread (Sullivan 2007; Morandini and Silvani 2010). The study in Sullivan (2007) considers a series of prescribed grassland fires conducted on flat terrain and finds no correlation between the measured rate of spread of the fire and N_C (note that under propagating fire conditions, the velocity u_w in Eq. (2.1) is replaced by the relative velocity $(u_w - \text{ROS})$). Similarly, the study in Morandini and Silvani (2010) considers a series of prescribed vegetation fires conducted on sloped terrain, identifies two different flame spread regimes, but finds no correlation between observed fire behavior and N_C. These results are difficult to interpret: data presented in Sullivan (2007) and Morandini and Silvani (2010) correspond to field-scale studies that feature propagating fires under variable wind conditions and over vegetation beds with finite thickness, and, in such complex systems, the effects of N_C could be masked by the effects of other parameters or by experimental uncertainties. In addition, the data analysis presented in Morandini and Silvani (2010) may be questionable because, as pointed out in Nelson, (2015), under sloped terrain conditions, the expression of N_C should be modified to include a correction for the slope angle.

Our own results, presented in Sections 2.3 and 2.4, confirm the importance of Byram's convection number for gas-fueled line flames exposed to wind. These results come from an integral model based on the classical theory of buoyant plumes in the presence of wind; the integral analysis suggests that the plume tilt angle is uniquely determined by N_C. The results also come from fine-grained Large Eddy Simulation (LES) performed for different values of the cross-wind velocity,

u_w, and different values of the fire intensity, $\dot{Q}'$; the LES results suggest that the transition between the detached and attached flame regimes occurs for values of N_C close to 1.

In the second configuration with slope (and without wind), the literature suggests that the flame structure is controlled by the slope angle, noted θ_s (θ_s is defined as the angle between the sloped surface and the horizontal plane, see Figure 2.2(c) and (d)). Low values of θ_s correspond to a plume-dominated fire; high values of θ_s correspond to a slope-driven fire. The literature suggests that the transition between the two flame regimes occurs for critical values of θ_s, noted $\theta_{s,critical}$, between 10 and 27 degrees (i.e. for values of the slope between 18 and 51%) (Drysdale and Macmillan 1992; Smith 1992; Woodburn and Drysdale 1998; Wu et al. 2000; Sharples et al. 2010). Similar to wind-driven fires, slope-driven fires are associated with an attached flame and plume geometry, and with large values of the rate of spread (values of ROS that are at least an order of magnitude larger than those observed in the plume-dominated regime). Slope-driven fires also feature strong unsteady effects and sustained flame acceleration, which suggests that, in the attached flame regime, ROS should be viewed as a time-dependent dynamical quantity (Viegas 2004).

The large variations in the observed values of the critical angle for transition between the detached and attached flame regimes are explained by the sensitivity of the line fire configuration to three-dimensional effects, in particular the presence or absence of lateral entrainment of air into the flame and plume regions. It is found that the transition to the attached flame regime is promoted by the presence of sidewalls that prevent lateral air entrainment, inhibit three-dimensional effects, and make the flame-flow configuration statistically two-dimensional: configurations with side walls correspond to lower values of $\theta_{s,critical}$ (Woodburn and Drysdale 1998; Sharples et al. 2010). These configurations are often referred to as "trenches" or "canyons" in the literature. In contrast, configurations without side walls feature strong edge effects, three-dimensional effects, and correspond to larger values of $\theta_{s,critical}$ (Dupuy et al. 2011; Huang and Gollner 2014).

The transition between the plume-dominated and slope-driven fire regimes is also associated with a change in the dominant mode of heat transfer that controls flame spread. It is found that, while radiation heat transfer tends to dominate under detached flame conditions, that is, at low values of θ_s, convection heat transfer tends to dominate under attached flame conditions, that is, at high values of θ_s (Dupuy and Maréchal 2011; Morandini et al. 2018). The dominant role of radiation and convection, at small and large values of the slope angle, respectively, can be explained in part by considering the flow direction in the vicinity of the flame: in the plume-dominated regime, the surface located upslope of the flame is exposed to downslope flow of ambient air, that is, to a "reverse" flow and convective cooling; in contrast, in the slope-driven regime, the upslope surface is exposed to upslope flow of combustion gases, that is, to convective heating. Note that one of the important features of line fires in the presence of wind or slope is the asymmetry of the air entrainment process into the flame and plume regions: the air entrainment capacity is higher on the upwind/downslope side of the fire and lower on the downwind/upslope side; this imbalance in the air entrainment capacity acts to accentuate the deflection of the flame and plume.

In summary, in configurations with wind, the structure of line fires is characterized in terms of Byram's convection number, N_C, while, in configurations with slope, the same structure is characterized in terms of the slope angle, θ_s. The question of how these two viewpoints can be reconciled is not addressed in the literature and remains an open question. A possible approach to reconcile the viewpoints based on N_C and θ_s can, however, be derived from the work presented in Nelson (2002): in Nelson (2002), the effects of slope are first represented through the introduction of an effective wind velocity, noted $u_{w,Eff}$; the velocity $u_{w,Eff}$ is then treated as the projection on the sloped surface of the characteristic velocity of the line fire in the vertical direction; and finally this characteristic velocity is taken as the reference buoyant velocity w_B, obtained in the absence of wind and slope. In the absence of wind, one writes (Nelson 2002):

$$u_{w,Eff} = w_B \times \sin(\theta_s) = \left[C_b \times \left(\frac{(1-\chi_{rad})\dot{Q}'g}{(\rho_a c_{p,a} T_a)} \right)^{(1/3)} \right] \times \sin(\theta_s), \quad (2.2)$$

where C_b is a model coefficient. This expression can now be substituted into Eq. (2.1) and thereby produce an estimate of Byram's convection number in the case of sloped terrain (and without wind):

$$N_C = \frac{2}{(C_b \times \sin(\theta_s))^3}. \quad (2.3)$$

This argument suggests that the transition between the detached and attached flame regimes is uniquely controlled by the slope angle θ_s and is thereby consistent with reported observations and implicit assumptions made in the literature. Our own results, presented in Sections 2.3 and 2.4, confirm the importance of the slope angle for gas-fueled line flames evolving along an inclined surface (and without lateral air entrainment). These results come from fine-grained LES performed for different values of the slope angle, θ_s; the LES results suggest that the transition between the detached and attached flame regimes occurs for values of θ_s close to 24 degrees.

2.3 Integral Model

We now turn to a brief description of the integral model used to develop a simple scaling analysis of the structure of line fires in wind-aided configurations (without slope). This model is based on the classical theory of weak buoyant plumes in the presence of wind and includes modifications introduced to describe strong plume effects (i.e. effects of large variations in mass density) as well as combustion effects in the near-field flame region. The model combines ideas developed in the atmospheric dispersion research community (Hoult et al. 1969; Krishnamurthy and Hall 1987), the wildland fire research community (Albini 1981; Mercer and Weber 1994; Nelson et al. 2012), and the combustion research community (Escudier 1972, 1975).

The model solves for the mean flame and plume structure through integral equations that express conservation of mass, vertical and horizontal momentum, heat, and fuel and oxygen mass. The model assumes steady state and top-hat profiles inside the flame and plume (all quantities are implicitly averaged in time and in space across the thickness of the flame/plume). The unit vector $\vec{s}$ of coordinates $(\cos(\theta), \sin(\theta))$ marks the orientation of the centerline of the flame and plume (see Figure 2.3). We write:

$$\frac{d}{ds}\left(\dot{m}'_p\right) = 2\rho_a v_e \tag{2.4}$$

$$\frac{d}{ds}\left(\dot{m}'_p u_p \sin\left(\theta\right)\right) = \left(\rho_a - \rho_p\right)2bg \tag{2.5}$$

$$\frac{d}{ds}\left(\dot{m}'_p u_p \cos\left(\theta\right)\right) = \frac{d}{ds}\left(\dot{m}'_p\right)u_w \tag{2.6}$$

$$\frac{d}{ds}\left(\dot{m}'_p c_p T_p\right) = \frac{d}{ds}\left(\dot{m}'_p\right)c_p T_a + (1 - \chi_{rad})\dot{\Omega}''_F \Delta H_F \tag{2.7}$$

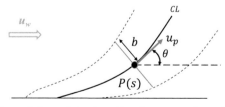

Figure 2.3 Integral model used to describe the mean flame/plume geometry in wind-aided configurations. The model calculates the variations of the half-thickness, b, inclination angle, θ, and relevant properties (e.g. the velocity u_p) as a function of the arc length, s, measured along the centerline of the flame and plume.

$$\frac{d}{ds}\left(\dot{m}'_p Y_{F,p}\right) = -\dot{\Omega}''_F \tag{2.8}$$

$$\frac{d}{ds}\left(\dot{m}'_p Y_{O_2,p}\right) = \frac{d}{ds}\left(\dot{m}'_p\right) Y_{O_2,a} - r_s \dot{\Omega}''_F, \tag{2.9}$$

where $\dot{m}'_p$ is the mass flow rate (per unit length of the fireline) in the direction $\vec{s}$, v_e is a characteristic velocity used to quantify air entrainment into the flame/plume, u_p the flow velocity in the direction $\vec{s}$, ρ_p the mass density, c_p the heat capacity at constant pressure (assumed constant), T_p the temperature, $\dot{\Omega}''_F$ the fuel mass reaction rate (defined as the volumetric fuel mass reaction rate averaged across the thickness of the flame/plume), ΔH_F the heat of combustion (per unit mass of fuel), $Y_{F,p}$ the fuel mass fraction, $Y_{O_2,p}$ the oxygen mass fraction, $Y_{O_2,a}$ the oxygen mass fraction in ambient air, and r_s the stoichiometric oxygen-to-fuel mass ratio in the assumed global combustion equation, $F + r_s O_2 \rightarrow (1 + r_s)P$ (1 kg of fuel F reacts with r_s kg of oxygen O_2 to yield $(1 + r_s)$ kg of products P). By definition, we have $\dot{m}'_p = \rho_p(2b)u_p$, with ρ_p related to T_p through the ideal gas law, $\rho_p \approx \rho_a\left(T_a/T_p\right)$.

The corresponding kinematic equations that give the spatial coordinates (x_c, z_c) of the point $P(s)$ along the centerline of the flame and plume are:

$$\frac{d}{ds}(x_c) = \cos(\theta), \tag{2.10}$$

$$\frac{d}{ds}(z_c) = \sin(\theta). \tag{2.11}$$

Equations (2.5) and (2.6) assume that the entrainment of ambient air into the flame and plume regions has no net effect on vertical momentum and has a net effect on horizontal momentum that is proportional to the wind velocity. Note that these assumptions are not consistent with the asymmetry of the air entrainment process noted in Section 2.2; this asymmetry is expected to result in a net negative force in the vertical direction as well as a net positive force in the downwind direction. These effects are simply neglected in the present formulation.

An alternative formulation for Eqs. (2.5) and (2.6) is often adopted in the literature:

$$\frac{d}{ds}\left(\dot{m}'_p u_p\right) = \frac{d}{ds}\left(\dot{m}'_p\right)(u_w \cos(\theta)) + \left(\rho_a - \rho_p\right) 2bg \sin(\theta), \tag{2.12}$$

$$\left(\dot{m}'_p u_p\right)\frac{d\theta}{ds} = -\frac{d}{ds}\left(\dot{m}'_p\right)(u_w \sin(\theta)) + \left(\rho_a - \rho_p\right) 2bg \cos(\theta). \tag{2.13}$$

Equations (2.4)–(2.9) or Eqs. (2.4), (2.7)–(2.9), (2.12), and (2.13) require a submodel for air entrainment and a submodel for combustion. The submodel for the turbulent air entrainment velocity is taken from Hoult et al. (1969):

$$v_e = \left(\alpha|u_p - u_w \cos(\theta)| + \beta|u_w \sin(\theta)|\right) \times \left(\frac{\rho_p}{\rho_a}\right)^{(1/2)}, \qquad (2.14)$$

where α and β are model coefficients taken from Mercer and Weber (1994): $\alpha = 0.16$ and $\beta = 0.5$. The expression in Eq. (2.14) assumes two mechanisms for air entrainment: a first mechanism due to the differential velocity between the plume and the ambient wind in the direction $\vec{s}$, $(u_p - u_w \cos(\theta))$; and a second mechanism due to the differential velocity between the plume and the ambient wind in the direction perpendicular to $\vec{s}$, $(u_w \sin(\theta))$. The last term in the expression for v_e, $(\rho_p/\rho_a)^{(1/2)}$ is a modification proposed in Escudier (1972) in order to account for strong density variations.

The submodel for combustion is similar to the formulation proposed in Escudier (1972, 1975):

$$\dot{\Omega}_F'' = GER \times \left(\frac{2\rho_a v_e \times Y_{O_2,a}}{r_s}\right), \qquad (2.15)$$

where GER is a model coefficient that represents a global equivalence ratio. The submodel assumes that the rate of combustion is mixing-limited and controlled by air entrainment: in Eq. (2.15), the product $(2\rho_a v_e \times Y_{O_2,a})$ is the local flux of oxygen mass entrained into the flame region; if all of this oxygen mass is consumed by combustion, we then have $\dot{\Omega}_F'' = (2\rho_a v_e \times Y_{O_2,a})/r_s$; and we thus see that in Eq. (2.15), GER simply measures the fraction of the entrained oxygen mass that is participating in the combustion process. The literature suggests that, under many conditions, the global equivalence ratio inside the flame region remains close to 0.1 and we therefore assume in the model that GER can be treated as a constant, $GER = 0.1$. Finally, it is worth noting that the combustion model in Eq. (2.15) is only applied in the flame region, identified by the condition $Y_F > 0$; when Y_F becomes negative, the model switches to the inert plume equations and uses $\dot{\Omega}_F'' = 0$.

Figure 2.4 presents representative results obtained with the integral model. The model is applied to a laboratory-scale configuration corresponding to a methane line burner characterized by a low value of the fire intensity, $\dot{Q}' = 100$ kW m^{-1}, and a small burner width, $w_{burner} = 5$ cm. The center of the burner is located at $(x, z) = (0, 0)$; the ground surface is horizontal, $\theta_s = 0$. Figure 2.4(a–c) presents results obtained with different values of the horizontal wind velocity, $u_w = 0.75$, 1.5, and 3 m s^{-1}, corresponding to different values of Byram's convection number, $N_C = 10.2$, 1.3, and 0.16, respectively. In the plots, the thick black solid line

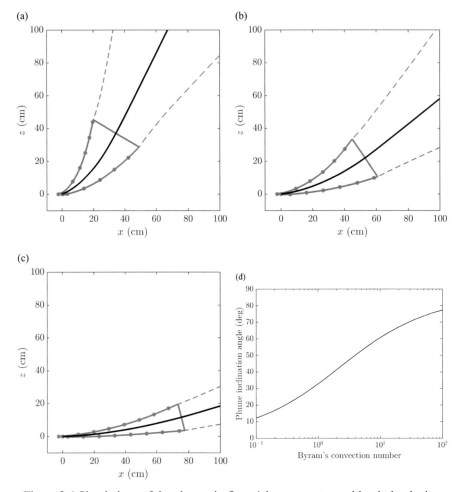

Figure 2.4 Simulations of the change in flame/plume structure with wind velocity, as predicted by the integral model. Flame/plume orientation and thickness simulated for different values of the crosswind velocity: (a) $u_w = 0.75$ m s^{-1}; (b) $u_w = 1.5$ m s^{-1}; (c) $u_w = 3$ m s^{-1}. (d): variations of the asymptotic value of the plume inclination angle, θ_∞, with Byram's convection number, N_C.

represents the location of the flame/plume centerline, that is, the location of the points P of coordinates (x_c, z_c); the red solid line represents the location of the edges of the flame region, defined as the region where $Y_F > 0$; the red dashed line represents the location of the edges of the plume region, defined as the region where $(T_p - T_a) > 50$ K. As expected, the model shows that at low velocities (i.e. at large values of N_C), the line fire features a vertically tilted geometry, while at high velocities (i.e. at low values of N_C), the fire features a horizontal geometry. Note that the integral model does not include ground surface effects and implicitly assumes that the flame and plume are evolving far from solid boundaries; this is clearly a serious limitation (and one that cannot be easily overcome) and, in its present form, the model is limited to qualitative use for physical insight and for the identification of controlling parameters.

One interesting result in Figure 2.4(a–c) is that, while the inclination angle θ varies significantly in the flame region, it is seen to achieve a constant asymptotic value in the plume region. We call θ_∞ this value and find that: for $u_w = 0.75$ m s^{-1}, $\theta_\infty \approx 61°$; for $u_w = 1.5$ m s^{-1}, $\theta_\infty \approx 36°$; and for $u_w = 1.5$ m s^{-1}, $\theta_\infty \approx 15°$. Based on this observation, one can develop a simplified analysis of the governing equations in Eqs. (2.4)–(2.9) valid in the far field region of the plume, where $\rho_p \approx \rho_a$, and derive the following solution:

$$b = \sin(\theta_\infty)(a\sin(\theta_\infty) + \beta\cos(\theta_\infty)) \times s \tag{2.16}$$

$$u_p = \frac{u_w}{\cos(\theta_\infty)} \tag{2.17}$$

$$(T_p - T_a) = \left(\frac{T_a u_w^2}{g}\right)\frac{\sin(\theta_\infty)}{\cos(\theta_\infty)^2} \times \frac{1}{s} \tag{2.18}$$

$$v_e = u_w \tan(\theta_\infty)(a\sin(\theta_\infty) + \beta\cos(\theta_\infty)) \tag{2.19}$$

$$\tan(\theta_\infty)^2(a\tan(\theta_\infty) + \beta) = \frac{N_C}{4}. \tag{2.20}$$

Thus, the integral model suggests that, in the far field region, the plume achieves a constant inclination angle (Eq. (2.20)) and a constant velocity (Eq. (2.17)), while the thickness increases linearly with arc length s (Eq. (2.16)) and the excess temperature decreases proportionally to the inverse of s (Eq. (2.18)). Equation (2.20) is the equation for the inclination angle, and this equation suggests that the far-field plume geometry is uniquely determined by Byram's convection number N_C. Figure 2.4(d) presents the corresponding variations of θ_∞ with N_C: the model predicts that $\theta_\infty \approx 45°$ for $N_C = 2.64$, which may be interpreted as an estimate of the critical value of N_C that delineates between the plume-dominated and wind-driven flame regimes.

In conclusion, consistent with much of the literature, the integral model suggests that Byram's convection number is the main parameter that controls the structure of wind-aided line fires. Some of the limitations of the integral model are noted: the asymmetry of the air entrainment process is neglected; the presence of solid boundaries is also neglected. Next, we explore further the same configuration and present a series of fine-grained LES, that is, simulations that are not limited by the simplifications made in the integral model.

2.4 Large Eddy Simulations

2.4.1 Numerical Solver

Numerical simulations presented in this section are performed using FireFOAM, a fire modeling solver developed by FM Global and based on an open-source, general-purpose, CFD software package called OpenFOAM[3]. FireFOAM is a second-order accurate, finite volume solver with implicit time integration; the solver features advanced meshing capabilities (structured or unstructured polyhedral mesh); it also features a massively parallel computing capability using Message Passing Interface protocols.

FireFOAM uses a Favre-filtered, compressible-flow, LES formulation, and provides a choice between several modeling options for the treatment of turbulence, combustion, and thermal radiation. In the baseline configuration: subgrid-scale (SGS) turbulence is described using the one-equation eddy viscosity model, which solves a transport equation for SGS turbulent kinetic energy (Fureby et al. 1997); combustion is described using the classical concept of a global combustion equation combined with the Eddy Dissipation Concept (EDC) model (Magnussen and Hjertager 1976); radiation is described by solving the Radiative Transfer Equation (RTE) using a discrete-ordinates, finite-volume method (Chai and Patanka 2006) and by assuming a nonscattering, nonabsorbing grey medium and using the concept of a prescribed global radiant fraction, χ_{rad}. This baseline configuration was adopted in our past work aimed at simulating an experimental turbulent line burner (Vilfayeau et al. 2016); we refer the reader to Vilfayeau et al. (2016) for details. The only difference between the present modeling choices and the previous baseline configuration is the substitution of the Wall-Adapting Local Eddy-viscosity (WALE) model (Nicoud and Ducros 1999) for the one-equation eddy viscosity model (for SGS turbulence); this model was previously adopted in Ren et al. (2016) and was shown to perform well in the vicinity of solid walls.

2.4.2 Numerical Configuration (with Wind and without Slope)

The numerical configuration is presented in Figure 2.5. The computational domain is 780-cm long in the streamwise x-direction, 50-cm wide in the spanwise y-direction and 250-cm high in the vertical z-direction. The line burner is 5-cm deep in the x-direction and 50-cm wide in the y-direction; the origin $(x, y, z) = (0, 0, 0)$ designates the y-center of the leading edge of the burner. The burner is flush-mounted on a 50-cm wide horizontal solid plate that extends from $x = (-20)$ cm to $x = 205$ cm. The air crossflow is injected through a 50-cm wide and 50-cm

---

[3] OpenFOAM, developed by the OpenFOAM Foundation, available at: www.openfoam.org (last accessed September 29, 2019).

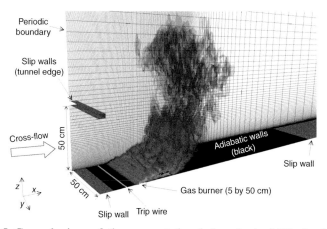

Figure 2.5 General view of the computational domain in LES simulations of the line fire configuration with wind. The burner is 5-cm deep in the x-direction and 50-cm wide in the y-direction; the wind tunnel at the inlet boundary of the computational domain is 50-cm wide in the y-direction and 50-cm high in the vertical z-direction. The flame is visualized using several isocontours of the instantaneous volumetric heat release rate.

A black and white version of this figure will appear in some formats. For the color version, please refer to the plate section.

high wind tunnel located at the inlet boundary of the computational domain, at $x = -30$ cm. Furthermore, a 5-mm deep, 50-cm wide, and 5-mm high trip wire is placed at $x = -10.5$ cm in order to perturb the incoming air flow and to thereby promote laminar-to-turbulent flow transition.

The computational mesh is a rectangular Cartesian grid. For $x \leq 100$ cm, grid spacing in the streamwise x-direction is uniform and equal to 5 mm; beyond that location, the x-grid is stretched with a stretch factor equal to 1.06. Grid spacing in the spanwise y-direction is uniform and is equal to 5 mm. This streamwise (spanwise) resolution corresponds to 10 (100) grid cells across the burner depth (width). Grid spacing in the vertical z-direction is non-uniform: the z-grid spacing is 1.2 mm at $z = 0$ (i.e. the first cell center is located 0.6 mm above the south boundary of the computational domain) and is 20 mm at $z = 50$ cm with a stretching factor of 1.04. For $z > 50$ cm, the z-grid is stretched with a stretch factor equal to 1.06. The total number of cells is 3.5 million.

These choices provide high levels of spatial resolution both in the flame region and in the boundary layer region close to the solid plate surface. The LES simulations are "wall-resolved" (Piomelli and Balaras 2002), that is, the near-wall grid spacing is sufficiently small to accurately capture the gradients of flow velocity at the plate surface and to calculate the wall shear stress directly from the LES solution, without the need for a subgrid-scale wall model. The only exception to the high spatial resolution requirement associated with wall-resolved LES is made in the trip wire region: the trip wire is under-resolved and is described with one grid cell in the x-direction and four grid cells in the z-direction.

The methane mass flow rate is prescribed at the injection boundary of the burner and the air velocity is prescribed at the vent boundary of the wind tunnel. The horizontal solid plate and the trip wire are both treated as no-slip adiabatic solid walls. The surface located at $z = 0$ between the wind tunnel vent and the leading edge of the solid plate is treated as a slip wall. The surface located at $z = 0$ beyond the solid plate, at $x > 205$ cm, is also treated as a slip wall. The side boundaries at $y = (-25)$ and 25 cm correspond to periodic boundary conditions. Other boundaries correspond to open flow conditions.

In all cases, the methane mass flow rate is linearly increased from 0 to 1 g s^{-1} during the first 5 seconds and is then held constant for the remainder of the simulations. This is done to allow the crossflow to establish itself over the line burner before the heat release rate of the fire reaches its nominal value of 50 kW. At nominal conditions, the fire intensity is $\dot{Q}' = 100$ kW m^{-1}. The crosswind velocity is varied between $u_w = 0.75$ and 3 m s^{-1}. Using $\chi_{rad} = 0.23$, these variations in u_w correspond to variations in Byram's convection number between $N_C = 10.2$ and 0.16. All simulations are performed for a duration of 30 seconds. Turbulent statistics are collected for the final 15 seconds of each simulation, after the flow and flame become statistically stationary and long enough for the statistics to be converged (to improve convergence, statistics are computed using both temporal- and spanwise-averaging). The time step is controlled by a classical Courant-Friedrichs-Lewy (CFL) condition, CFL $= 0.5$, and is approximately equal to 0.35 ms. Each simulation is run using 200 processors on a large-scale Linux cluster, with a typical simulation requiring 40,000 CPU-hours.

2.4.3 LES Results (Configuration with Wind and without Slope)

Figure 2.6 presents representative results obtained with LES in the form of instantaneous snapshots of the structure of the plume, defined as the region where $(\tilde{T} - T_a) > 100$ K. Note that the downstream region in Figure 2.6(c) is affected by reduced resolution due to stretching of the x-grid and should be ignored. As expected, the LES simulations show that at low velocities (that is at $u_w = 0.75$ m s^{-1}), the line fire features a vertically tilted, detached geometry, while at high velocities (i.e. at $u_w = 3$ m s^{-1}), the fire features a horizontal, attached geometry. In addition, Figure 2.6 shows that as u_w is increased, the turbulent plume evolves from a transitional structure dominated by large-scale two-dimensional motions to a fully-developed turbulent structure dominated by small-scale three-dimensional motions.

(a) (b)

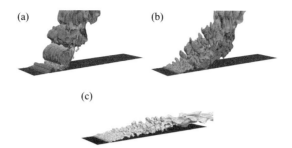

(c)

Figure 2.6 LES simulations of the change in flame/plume structure with wind velocity. Three-dimensional instantaneous structure of the plume for different values of the crosswind velocity: (a) $u_w = 0.75$ m s^{-1}; (b) $u_w = 1.5$ m s^{-1}; (c) $u_w = 3$ m s^{-1}. The plume is colored according to the local value of the x-velocity; red (white) corresponds to a low (high) value.
A black and white version of this figure will appear in some formats. For the color version, please refer to the plate section.

Figure 2.7 presents additional results in the form of instantaneous snapshots of the spatial variations of the flow velocity vector and of the flame and plume contours in the central vertical plane of the computational domain; the flame boundary is visualized using an isocontour of the volumetric heat release rate (50 kW m^{-3}); the plume boundary is visualized using an isocontour of temperature $((\tilde{T} - T_a) = 100$ K$)$. Note that for ease of visualization, only a subset of the simulated velocity vectors is being plotted in Figure 2.7. For $u_w = 0.75$ m s^{-1}, the flow is clearly dominated by the buoyancy of the flame: the cross-flow is strongly deflected in the upward direction; the flow along the wall surface (at $z = 0$) in the region located downwind of the flame is opposed to the wind flow direction, and air entrainment is therefore two-sided. Also, the end of the flame attachment region corresponds to the location where the wall flow separates from the surface (i.e. where the wall boundary layer reverses in flow direction). In contrast, for $u_w = 3.0$ m s^{-1}, the flow is now dominated by the momentum of the cross-wind: the cross-flow is only weakly deflected in the upward direction; the flow along the wall surface (at $z = 0$) in the downwind region of the flame is in the wind flow direction and air entrainment is therefore one-sided. As mentioned in Section 2.2, these differences in flow pattern are believed to be key ingredients in flame spread mechanisms (Dold and Zinoviev 2009; Finney et al. 2013, 2015).

Furthermore, a careful observation of the flow pattern presented in Figure 2.7(a) reveals some of the deficiencies in the assumptions made by the integral model discussed in Section 2.3. In the near-field region of the flame, that is, at $x < 30$ cm, the flame is clearly strongly affected by the presence of the solid surface and remains attached to that surface. In the far-field region, that is, at $30 \leq x \leq 50$ cm,

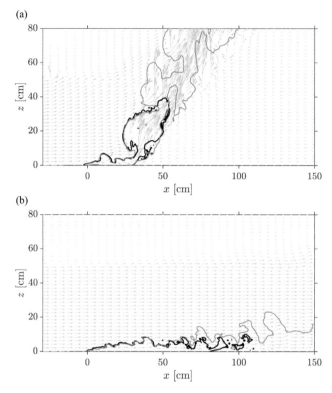

(a)

(b)

Figure 2.7 LES simulations of the change in flame/plume structure with wind velocity. Two-dimensional instantaneous structure of the flame (solid black line) and plume (solid grey line) for different values of the crosswind velocity: (a) $u_w = 0.75$ m s^{-1}; (b) $u_w = 3$ m s^{-1}.

the flame is detached from the surface but the flow fields upwind and downwind of the flame are seen to be significantly different and while air entrainment is two-sided, it is likely to be asymmetric with stronger entrainment expected on the upwind side. These observations suggest that the integral model presented in Eqs. (2.4)–(2.9) probably underestimates the deflection of the flame/plume due to wind.

While instantaneous snapshots reveal the rich dynamical content of the LES simulations, relevant information is also obtained by adopting a statistical viewpoint and by analyzing the mean structure of the flame and plume. Figure 2.8 presents the mean (that is, time- and spanwise-averaged) shape and location of the flame and plume, visualized using an isocontour of the mean volumetric heat release rate (50 kW m^{-3}) and an isocontour of the mean temperature $((\langle \tilde{T} \rangle - T_a) = 100$ K$)$, respectively. As u_w increases from 0.75 to 3.0 m s^{-1}, the flame geometry

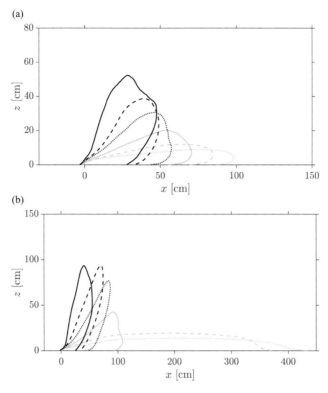

Figure 2.8 LES simulations of the change in flame/plume structure with wind velocity. Two-dimensional mean structure of (a) the flame and (b) the plume. From left to right: $u_w = 0.75$, 1.0, 1.25, 1.5, 2.0, 3.0 m s^{-1}.

(Figure 2.8(a)) is seen to evolve from a detached flame regime to an attached flame regime. The vertical elevation of the flame decreases (from 50 to 10 cm); the flame length (loosely defined as the distance from the burner to the flame tip) increases (from 50 to 100 cm); and the flame attachment length (defined as the x-wall-distance downstream of the burner in contact with the flame) increases (from 30 to 90 cm). In the present configuration, the transition from a vertically tilted flame to a horizontal flame is gradual and occurs between $u_w = 1.0$ and 1.5 m s^{-1}. Similar variations are observed in the mean plume shape (Figure 2.8(b)). As u_w increases from 0.75 to 3.0 m s^{-1}, the vertical elevation of the plume decreases (from 90 to 20 cm); the plume length (loosely defined as the distance from the burner to the plume tip) increases (from 90 to 400 cm; and the plume attachment length (defined as the x-wall-distance downstream of the burner in contact with the plume) increases (from 30 to 400 cm). The transition from a vertically tilted plume to an elongated horizontal plume is abrupt and occurs between $u_w = 1.5$ and 2.0 m s^{-1}.

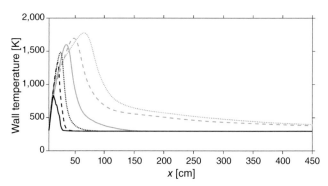

Figure 2.9 LES simulations of the change in flame/plume structure with wind velocity. Variations of the mean wall temperature with spatial distance. From left to right: $u_w = 0.75$, 1.0, 1.25, 1.5, 2.0, 3.0 m s^{-1}.

Additional information on the flame and plume attachment behavior is presented in Figure 2.9, which plots the spatial variations of the mean temperature along the (adiabatic) wall surface. In this figure, the flame attachment region can be identified as the region where the mean wall temperature increases; the plume attachment region is the region located downstream, beyond the point where the mean wall temperature reaches a maximum value. As mentioned in the discussion of Figure 2.8, as u_w increases from 0.75 to 3.0 m s^{-1}, the simulations predict a gradual increase in the flame attachment length as well as an abrupt increase in the plume attachment length for u_w between 1.5 and 2.0 m s^{-1}. The simulations also predict a gradual increase in the peak value of the mean wall temperature. Assuming that convection (rather than radiation) is the dominant heat transfer mechanism, the length of the region where the mean wall temperature is higher than 600 K (taken here as a representative value for the start of pyrolysis in typical vegetation fuel) can be interpreted as giving an estimate of the length of the preheating region in a flame spread scenario.

The LES results can be compared to the predictions of the integral model discussed in Section 2.3. For comparison purposes, the mean plume shape simulated by LES is characterized as follows (Figure 2.10): first the plume attachment length, $x_{p,a}$, is calculated as the x-wall-distance downstream of the burner in contact with the plume (where $(\langle \tilde{T} \rangle - T_a) > 100$ K); point B is then defined as the mid-point of the base of the plume at $z = 0$, that is, as the point of coordinates $(x, z) = (0.5 \times x_{p,a}, 0)$; next, the tip point T is defined as the point on the plume edge contour that is located at the maximum distance from B (note that this definition is preferred to a definition of T as the point on the plume edge contour with maximum elevation; the two definitions are similar in the case of detached plumes but the present definition provides a better description of the

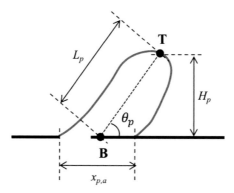

Figure 2.10 Analysis of the mean plume shape simulated by LES. Point B is the mid-point of the base of the plume; point T is the tip of the plume; θ_p is the plume inclination angle.

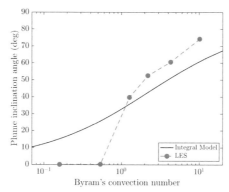

Figure 2.11 Variations of the plume inclination angle, θ_p, with Byram's convection number, N_C. Comparison between results obtained with the integral model (solid line) and with LES (dashed line with dot symbols).

plume shape in the case of attached plumes); the line that connects B to T allows us to define the plume length, L_p, and plume vertical height, H_p; finally, the plume inclination angle θ_p is obtained as the angle between the B–T line and the horizontal plane.

Figure 2.11 presents the variations of the plume inclination angle, θ_p, with the crosswind velocity, u_w, re-cast in the form of Byram's convection number N_C. The figure compares predictions obtained with the integral model discussed in Section 2.3 and with LES. The LES data suggests that the transition between the attached plume regime ($\theta_p = 0$) and the detached plume regime ($\theta_p > 0$) is abrupt and occurs for a value of N_C between 0.5 ($u_w = 2$ m s^{-1}) and 1.3 ($u_w = 1.5$ m s^{-1}). In the detached plume regime ($N_C \geq 1.3$, $u_w \leq 1.5$ m s^{-1}), the comparison between predictions by the integral model and by LES is relatively good: the values of θ_p obtained in the integral model are within 10 degrees of the LES values. The integral model, however, does not capture the transition to the attached plume regime observed in the LES at low values of N_C ($N_C \leq 0.5, u_w \geq 2$ m s^{-1}). Recall that these results are obtained for a fixed value of the fire intensity, $\dot{Q}' = 100$ kW m^{-1}; additional simulations (not shown here) performed with different values of the fire intensity, $\dot{Q}' = 20$ and 500 kW m^{-1}, confirm that the transition between the plume-dominated and wind-driven line fires occurs for a critical value of Byram's convection number between 0.5 and 2.

2.4.4 Numerical Configuration (with Slope and without Wind)

The numerical configuration is presented in Figure 2.12. The computational domain is first constructed at zero slope angle with the same grid spacing and dimensions already used in the wind-aided configuration, except that the burner location, that is, the origin $(x, y, z) = (0, 0, 0)$, is now placed at equal distance from the x-boundaries, and that the trip wire and the wind tunnel are removed. The domain is then tilted to model different slope angles while keeping the grid lines normal or parallel to the inclined surface. The heat release rate of the fire is 50 kW; the fire intensity is $\dot{Q}' = 100$ kW m^{-1}.

Figure 2.12 General view of the computational domain in LES simulations of the line fire configuration with slope. The computational mesh is tilted so that grid lines remain normal or parallel to the inclined surface. The flame is visualized using volume rendering of temperature.
A black and white version of this figure will appear in some formats. For the color version, please refer to the plate section.

2.4.5 LES Results (Configuration with Slope and without Wind)

Figure 2.13 presents instantaneous snapshots of the spatial variations of the flow velocity vector plotted with the flame and plume contours in the central vertical plane of the computational domain; the flame and plume contours are visualized using the same methodology already used in Figure 2.7. For $\theta_s = 18$ degrees, the flow is clearly dominated by vertical motions: the flame and plume are detached from the surface; the flow direction upslope of the flame is opposed to the flow direction downslope of the flame and air entrainment is two-sided. In contrast, for $\theta_s = 36$ degrees, the flow is now dominated by strong upslope motions: the flame and plume feature a boundary layer geometry; the flow along the wall surface

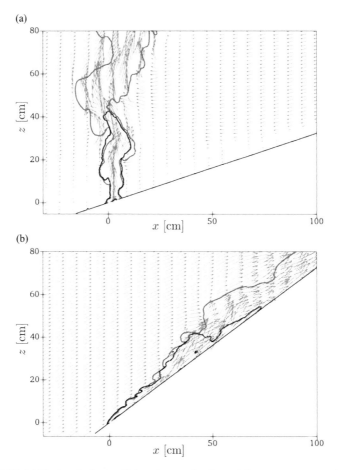

Figure 2.13 LES simulations of the change in flame/plume structure with slope angle. Two-dimensional instantaneous structure of the flame (solid black line) and plume (solid grey line) for different values of the slope angle: (a) $\theta_s = 18$ degrees; (b) $\theta_s = 36$ degrees.

remains in the upslope direction and air entrainment is therefore one-sided. The strong difference in the air entrainment process observed when comparing Figures 2.13(a) and 2.13(b) illustrates the concept of an effective wind created by the line fire when developing along an inclined surface: while for $\theta_s = 18$ degrees, the mean entrainment velocity measured downslope of the flame is on the order of 0.3 m s^{-1}, this velocity is more than doubled for $\theta_s = 36$ degrees. This increase in the downslope entrainment velocity is associated with, and can be explained by, the effects of wall blockage and the corresponding reduction in entrainment velocity occurring on the upslope side of the flame.

Figure 2.14 presents the mean (i.e. time- and spanwise-averaged) shape and location of the flame and plume; the mean flame and plume contours are visualized using the same methodology already used in Figure 2.8; in contrast to the choice made in Figure 2.13, the contours are now presented in a tilted plane attached to the inclined surface. As θ_s increases from 9 to 36 degrees, the flame geometry (Figure 2.14(a)) is seen to evolve from a detached flame regime to an attached flame regime. The normal (i.e. normal to the wall surface) elevation of the flame decreases (from 60 to 10 cm); the flame length increases (from 60 to 100 cm); and

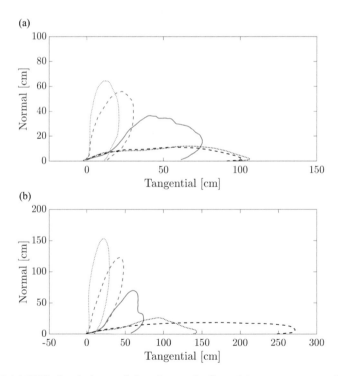

Figure 2.14 LES simulations of the change in flame/plume structure with slope angle. Two-dimensional mean structure of (a) the flame and (b) the plume. From left to right: $\theta_s = 9, 18, 24, 30, 36$ degrees.

the flame attachment length increases (from 10 to 100 cm). In the present configuration, the transition from a vertically tilted flame to a wall flame is abrupt and occurs between $\theta_s = 18$ and 30 degrees, approximately at $\theta_s \approx 24$ degrees. Similar variations are observed in the mean plume shape (Figure 2.14(b)). As θ_s increases from 9 to 36 degrees, the normal elevation of the plume decreases (from 150 to 20 cm); the plume length increases (from 150 to 270 cm); and the plume attachment length increases (from 10 to 250 cm). The transition from a vertically tilted plume to an elongated wall plume is abrupt and occurs approximately at $\theta_s \approx 24$ degrees.

2.5 Conclusion

Fine-grained LES are presented in this chapter to bring fundamental insight into the effects of crosswind or sloped terrain on the structure of buoyancy-driven, turbulent line fires. The LES simulations describe the transition between a pool-like, detached flame geometry, characterized by a tilted vertical shape, and a boundary layer, attached flame geometry, characterized by a wall-attached shape. The detached flames feature downwind/upslope flow separation and two-sided air entrainment into the flame. In contrast, the attached flames feature unidirectional flow and one-sided air entrainment into the flame. The analysis of the LES results is supported by an integral model that helps identify the dominant parameters that control the transition between the detached and attached flame/plume regimes.

The present flame configurations are simplified: the flames are gas-fueled, correspond to a small fixed value of the burner width, and are statistically two-dimensional (i.e. they correspond to line fires developing in "trenches" or in "canyons"); in addition, the flames correspond to weak-to-moderate fires (the fire intensities are below 500 kW m^{-1}). In these simplified configurations, the transition between the two flame regimes is observed for a critical value of Byram's convection number N_C close to 1 in the case of wind-aided flames, and for a critical value of the slope angle θ_s close to 24 degrees in the case of slope-aided flames. There is a need to confirm the present results in configurations featuring a range of values of the burner width as well as larger values of the fire intensity.

Future research should also be directed at extending the reach of detailed analysis to the case of flames propagating over representative vegetation beds. One objective is to bring higher fidelity to the geometric description of the fuel sources, in particular through a treatment of the vegetation fuel as a discontinuous system with effects associated with fuel particle size distributions and separation distances. Another objective is to bring higher fidelity to the description of flame spread dynamics through a treatment of the uncontrolled production of flammable vapors due to pyrolysis processes and to the gas-to-solid heat transfer as well as a treatment of the typical cycle of ignition, pyrolysis and end-of-pyrolysis/burn-out. And finally, a related objective is to bring higher fidelity to the description of flame spread dynamics through a treatment of the displacement of the flame.

References

Albini, FA (1981) A model for the wind-blown flame from a line fire. *Combustion and Flame* **43**, 155–174.

Canfield, JM, Linn, RR, Sauer, JA, Finney, M, Forthofer, J (2014) A numerical investigation of the interplay between fireline length, geometry, and rate of spread. *Agricultural and Forest Meteorology* **189–190**, 48–49.

Chai, J, Patanka, S (2006) Discrete-ordinates and finite-volume methods for radiation heat transfer. In: Minkowycz, WJ, Sparrow, EM, Murthy, JY, eds. *Handbook of Numerical Heat Transfer*. Hoboken, NJ: John Wiley & Sons, pp. 297–323.

Clark, TL, Coen, J, Latham, D (2004) Description of a coupled atmosphere–fire model. *International Journal of Wildland Fire* **13**(1), 49–64.

Coen, JL, Cameron, M, Michalakes, J, Patton, EG, Riggan, PJ, Yedinak, KM. (2013) Wrf-fire: Coupled weather–wildland fire modeling with the weather research and forecasting model. *Journal of Applied Meteorology and Climatology* **52**(1), 16–38.

Dold, JW, Zinoviev, A (2009) Fire eruption through intensity and spread rate interaction mediated by flow attachment. *Combustion Theory and Modelling* **13**(5), 763–793.

Drysdale, DD, Macmillan, AJR (1992) Flame spread on inclined surfaces. *Fire Safety Journal* **18**, 245–254.

Dupuy, J-L, Maréchal, J (2011) Slope effect on laboratory fire spread: Contribution of radiation and convection to fuel bed preheating. *International Journal of Wildland Fire* **20**(2), 289–307.

Dupuy, J-L, Maréchal, J, Portier, D, Valette, J-C (2011) The effects of slope and fuel bed width on laboratory fire behaviour. *International Journal of Wildland Fire* **20**(2), 272–288.

Escudier, MP (1972) Aerodynamics of a burning turbulent gas jet in a crossflow. *Combustion Science and Technology* **4**(1), 293–301.

Escudier, MP (1975) Analysis and observations of inclined turbulent flame plumes. *Combustion Science and Technology* **10**(3–4), 163–171.

Filippi, JB, Bosseur, F, Mari, C, Lac, C, Moigne, PL, Cuenot, B, Veynante, D, Cariolle, D, Balbi, J (2009) Coupled atmosphere–wildland fire modelling. *Journal of Advances in Modeling Earth Systems* **1**(4), 11.

Filippi, JB, Pialat, X, Clements, CB (2013) Assesment of forefire/meso-nh for wildland fire/atmosphere coupled simulation of the fireflux experiment. *Proceedings of the Combustion Institute* **34**(2), 2633–2640.

Finney, MA, Cohen, JD, Forthofer, JM, McAllister, SS, Gollner, MJ, Gorham, DJ, Saito, K, Akafuah, NK, Adam, BA, English, JD. (2015) Role of buoyant flame dynamics in wildfire spread. *Proceedings of the National Academy of Sciences (USA)* **112**(32), 9833–9838.

Finney, MA, Cohen, JD, McAllister, SS, Jolly, WM (2013) On the need for a theory of wildland fire spread. *International Journal of Wildland Fire* **22**(1), 25–36.

Fureby, C, Tabor, G, Weller, HG, Gosman, AD (1997) A comparative study of subgrid scale models in homogeneous isotropic turbulence. *Physics of Fluids* **9**(5), 1416–1429.

Hoult, DP, Fay, JA, Forney, LJ (1969) A theory of plume rise compared with field observations. *Journal of the Air Pollution Control Association* **19**(8), 585–590.

Hu, L (2017) A review of physics and correlations of pool fire behaviour in wind and future challenges. *Fire Safety Journal* **91**, 41–55.

Huang, X, Gollner, MJ (2014) Correlations for evaluation of flame spread over an inclined surface. *Proceedings of the Eleventh International Symposium, International Association for Fire Safety Science (IAFSS)* **13**, 222–233.

Kochanski, AK, Jenkins, MA, Mandel, J, Beezley, JD, Clements, CB, Krueger, S (2013) Evaluation of wrf-sfire performance with field observations from the fireflux experiment. *Geoscientific Model Development* **6**(4), 1109–1126.

Krishnamurthy, R, Hall, JD (1987) Numerical and approximate solutions for plume rise. *Atmospheric Environment* **21**(10), 2083–2089.

Lam, CS, Weckman, EJ (2015a) Wind-blown pool fire, Part I: Experimental characterization of the thermal field. *Fire Safety Journal* **75**, 1–13.

Lam, CS, Weckman, EJ (2015b) Wind-blown pool fire, Part II: Comparison of measured flame geometry with semi-empirical correlations. *Fire Safety Journal* **78**, 130–141.

Linn, RR, Cunningham, P (2005) Numerical simulations of grass fires using a coupled atmosphere-fire model: Basic fire behavior and dependence on wind speed. *Journal of Geophysical Research* **110**, D13107.

Magnussen, BF, Hjertager, BH (1976) On mathematical modeling of turbulent combustion with special emphasis on soot formation and combustion. *Symposium (International) on Combustion* **16**(1), 719–729.

Mandel, J, Beezley, JD, Kochanski, AK (2011) Coupled atmosphere–wildland fire modeling with wrf 3.3 and sfire 2011. *Geoscientific Model Development* **4**(3), 591–610.

Mell, W, Jenkins, MA, Gould, J, Cheney, P (2007) A physics-based approach to modeling grassland fires. *International Journal of Wildland Fire* **16**(1), 1–22.

Mercer, GN, Weber, RO (1994) Plumes above line fires in a cross wind. *International Journal of Wildland Fire* **4**(4), 201–207.

Morandini, F, Silvani, X (2010) Experimental investigation of the physical mechanisms governing the spread of wildfires. *International Journal of Wildland Fire* **19**(5), 570–582.

Morandini, F, Silvani, X, Dupuy, J-L, Susset, A (2018) Fire spread across a sloping fuel bed: Flame dynamics and heat transfers. *Combustion and Flame* **190**, 158–170.

Morvan, D (2011) Physical phenomena and length scales governing the behavior of wildfires: A case for physical modelling. *Fire Technology* **47**(2), 437–460.

Morvan, D, Dupuy, JL (2001) Modeling of fire spread through a forest fuel bed using a multiphase formulation. *Combustion and Flame* **127**(1–2), 1981–1994.

Morvan, D, Dupuy, JL (2004) Modeling the propagation of a wildfire through a mediterranean shrub using a multiphase formulation, *Combustion and Flame* **138**(3), 199–210.

Morvan, D, Frangieh, N (2008) Wildland fires behavior: Wind effect versus Byram's convective number and consequences upon the regime of propagation. *International Journal of Wildland Fire* **27**(9), 636–641.

Morvan, D, Porterie, B, Larini, M, Loraud, JC. (1998) Numerical simulation of turbulent diffusion flame in cross flow. *Combustion Science and Technology* **140**(1–6), 93–122.

Nelson, RM (1993) Byram's derivation of the energy criterion for forest and wildland fires. *International Journal of Wildland Fire* **3**(3), 131–138.

Nelson, RM (2002) An effective wind speed for models of fire spread. *International Journal of Wildland Fire* **11**(2), 153–161.

Nelson, RM (2015) Re-analysis of wind and slope effects on flame characteristics of Mediterranean shrub fires. *International Journal of Wildland Fire* **24**(7), 1001–1007.

Nelson, RM, Butler, BW, Weise, DR (2012) Entrainment regimes and flame characteristics of wildland fires. *International Journal of Wildland Fire* **21**(2), 127–140.

Nicoud, F, Ducros, F (1999) Subgrid-scale stress modelling based on the square of the velocity gradient tensor. *Flow, Turbulence and Combustion* **62**(3), 183–200.

Nmira, F, Consalvi, JL, Boulet, P, Porterie, B (2010) Numerical study of wind effects on the characteristics of flames from non-propagating vegetation fires. *Fire Safety Journal* **45**(2), 129–141.

Pagni, PJ, Peterson, TG (1973) Flame spread through porous fuels. *Symposium (International) on Combustion* **14**(1), 1099–1107.

Piomelli, U, Balaras, E (2002) Wall-layer models for large-eddy simulations. *Annual Review of Fluid Mechanics* **34**, 349–374.

Porterie, B, Morvan, D, Loraud, JC, Larini, M (2000) Firespread through fuel beds: Modeling of wind-aided fires and induced hydrodynamics. *Physics of Fluids* **12**(7), 1762–1782.

Putnam, AA (1965) A model study of wind-blown free-burning fires. *Symposium (International) on Combustion* **10**(1), 1039–1046.

Raupach, MR (1990) Similarity analysis of the interactions of bushfire plumes with ambient winds. *Mathematical and Computer Modelling* **13**(12), 113–121.

Ren, N, Wang, Y, Vilfayeau, S, Trouvé, A (2016) Large eddy simulation of turbulent vertical wall fires supplied with gaseous fuel through porous burners. *Combustion and Flame* **169**(3), 194–208.

Sharples, JJ, Gill, AM, Dold, JW (2010) The trench effect and eruptive wildfires: Lessons from the King's Cross underground disaster. *Proceedings of the Australasian Fire and Emergency Service Authorities Council (AFAC) Conference*, September, Darwin.

Smith, DA (1992) Measurements of flame length and flame angle in an inclined trench. *Fire Safety Journal* **18**(3), 231–244.

Sullivan, AL (2007) Convective froude number and Byram's energy criterion of Australian experimental grassland fires. *Proceedings of the Combustion Institute* **31**(2), 2557–2564.

Tang, W, Miller, CH, Gollner, MJ (2017) Local flame attachment and heat fluxes in wind-driven line fires. *Proceedings of the Combustion Institute* **36**(2), 3253–3261.

Viegas, DX (2004) On the existence of a steady state regime for slope and wind driven fires. *International Journal of Wildland Fire* **13**(1), 101–117.

Viegas, DX (2005) A mathematical model for forest fires blowup. *Combustion Science and Technology* **177**(1), 27–51.

Vilfayeau, S, White, JP, Sunderland, PB, Marshall, AW, Trouvé, A (2016) Large eddy simulation of flame extinction in a turbulent line fire exposed to air–nitrogen co-flow. *Fire Safety Journal* **86**, 16–31.

Woodburn, PJ, Drysdale, DD (1998) Fires in inclined trenches: The dependence of the critical angle on the trench and burner geometry. *Fire Safety Journal* **31**, 143–164.

Wu, Y, Xing, HJ, Atkinson, G (2000) Interaction of fire plume with inclined surface. *Fire Safety Journal* **35**(4), 391–403.

Yuan, L-M, Cox, G (1996) An experimental study of some line fires. *Fire Safety Journal* **27**(2), 123–139.

3

Energy Transport and Measurements in Wildland and Prescribed Fires

BRET W. BUTLER AND JOSEPH J. O'BRIEN

3.1 Introduction

In a set of experiments two different fuels were exposed to various levels of radiant heating (Cohen and Finney 2010; Finney et al. 2010). The first was a mass (about a baseball size) of shredded Aspen wood shavings, the second was a small solid pine wood block approximately 2 cm by 1 cm by 6 cm. The question posed was which would ignite first under a known constant radiant heat load at a magnitude that can result in ignition? Intuitively, many persons would say that the mass of fine wood shavings would be the first to ignite, but the opposite was true. The underlying explanation being that as the fine wood shavings were heated by radiant energy heating, their surface temperature began to rise above the surrounding air. The air near the wood shaving surface was then heated and formed buoyant plumes. The natural convection due to these small microplumes resulted in convective cooling of the wood particle surfaces that offset the radiant heating. For the small wood block sufficient natural convection could not form and ignition occurred much sooner than for the mass of wood shavings. This experiment illustrates the importance of considering both fuel and environment properties in the context of convective and radiant heating in nearly all cases of energy transport in fires.

The ignition and subsequent spread of fire is a complex process driven by interactions between energy transport, mass transport, and chemical kinetics. While the focus of this chapter is energy transport and how to capture it, a basic understanding of the concepts within the context of wildland fire requires an underlying knowledge of the fundamental processes of pyrolysis, ignition and sustained burning (see Chapter 1). The approach followed for this chapter is first to present a basic description of heat transfer mechanisms in the context of flames and the importance of quantifying these processes to explain wildland fire phenomena. Then, we present a fundamental description of the pyrolysis and ignition processes with a description of what constitutes a flame. We end with a presentation of basic magnitudes of heating that have been reported from wildland flames, along with some guidance on their measurement, and summarize fundamental research needs.

3.2 Heat Transfer

Heat transfer occurs through three modes: (1) conduction – which is energy transport from one solid to another solid that is in direct contact; (2) convection – which typically is energy transport from a fluid or gas to a solid; and (3) radiation – which is transported typically from a solid or gas to another entity through space by electromagnetic waves of radiation, with a concentration of energy in the infrared portion of the spectrum, that otherwise behaves much like visible light (Figure 3.1). Radiative and convective heat transfer have been determined to dominate energy transport in wildland fires (Anderson 1969; Butler et al. 2004; Yedinak et al. 2006; Anderson et al. 2010; Frankman et al. 2012).

Historically it has been assumed that, at least for crown fires, radiant energy transport dominates the energy exchange process (Albini 1986). However, recent measurements have shown that both radiation and convection are critical to wildland fire propagation. Anecdotal observations suggest some cases exist where radiation indeed dominates the energy transport process, for example a fire spreading through grass in the absence of wind would seem to be driven by radiant heating ahead of the flaming front, or a large crown fire with minimal driving wind would also be characterized by primarily radiant heating, although in both cases it is difficult to separate the radiant heating from the ignition of spot fires ahead of the main fire front due to lofting of embers that act as ignition sources and the flame movement driven by turbulence and entrainment. The opposite limit (primarily convective driven flames) would be a fire burning through grass in the presence of a very strong ambient wind. The wind causes the flames to tilt forward and reach far ahead of the burning front, effectively bathing the fuels in hot gases and convectively preheating vegetation far in advance of the fire. Intuitively, in this case convective energy transport would be significant and potentially dominate energy transport.

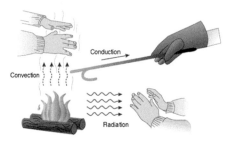

Figure 3.1 General relation between energy transfer modes.

A set of experiments conducted in a controlled laboratory environment (Anderson et al. 2010; Frankman et al. 2010; Yedinak et al. 2010) suggest that heating from wildland flames can be divided into roughly two regimes: (1) the first regime is characterized as heating that occurs far from the flame (i.e. greater than one flame length) and is dominated by radiant energy transport and (2) as the flame approaches within one flame length and closer a second regime develops where radiant energy transport increases exponentially, but convective energy transport is significant, with peak heating values exceeding the radiant heat peaks, sometimes by twice as much (Frankman et al. 2012). In all cases the ignition occurs through piloted ignition, which by definition is ignition of a combustible gas mixture by a spark or small flame. At some distance (nominally greater that a flame length) convection generally acts in a cooling mode, counteracting the radiant driven heating; however, just prior to arrival of the flame front, convective heating reverses and contributes substantial energy to the heating process (Anderson et al. 2010) and is critical to piloted ignition of the vegetation. A current focus of wildland fire research is to define the relative contribution of radiative and convective energy transport in diverse combustion scenarios that represent wildland fires and their interactions to varying degrees.

This chapter discusses the basic concepts of radiative and convective heating, their fundamental characteristics, the current theory on their role in wildland fire, and some techniques for their measurement.

3.2.1 Radiant Energy Transport

A person standing around a campfire would feel primarily radiative energy until they held their hand directly above the flame, at which time they would experience primarily convective heating on their hand. Planck's law states that radiant energy intensity is proportional to the temperature of the flame raised to the power of 4 multiplied by emissivity of the source: $q = \varepsilon \alpha T^4$, where q is radiant heat flux, ε is the emitting surface emissivity which ranges from 0 to 1 (for most natural biomass it is in the range of 0.7–0.95), α is the Stefan–Boltzmann constant (5.67×10^{-8} W m^{-2} K^{-4}), and T is the temperature of the emitting surface (Figure 3.2; Plank's Law requires units of absolute temperature K, whereas most measurements are reported in °C = K – 273.15. Measurements in some disciplines and industrial settings are still reported in °F). For soot-laden flames, much like those found in wildland fires, emissivity is generally a constant that is near 1. It can be calculated from $\varepsilon = 1 - e^{-kl}$, where k is the absorption coefficient and depends on wavelength of the energy. For large kl, ε approaches unity. High emissivity surfaces will more efficiently radiate energy than low emissivity surfaces.

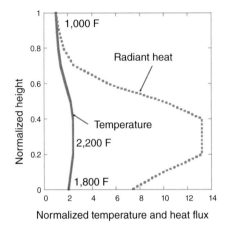

Figure 3.2 Relation between flame temperature and emitted radiant energy normalized by measured height and 1,000°F. Note that a slight increase in temperature produces a large increase in radiant heat.

Due to the strong temperature dependence, seemingly small changes in temperature of the hot surface can have significant impacts on the total energy emitted. For example, a common temperature in wildland flames is 1,000°C, which, assuming an emissivity of 1, results in an emitted radiant heat flux of approximately 149 kW m^{-2}. An increase to 1,200°C results in an emitted flux of approximately 266 kW m^{-2}, a nearly 80% increase in emitted energy for a 20% change in temperature. Thus, small increases in flame temperature result in large increases in radiated energy. However, distance from the source of heat also plays a factor in the heating intensity. Intensity decreases with the inverse square of the distance from the flames.

Raj (2008) suggests that absorption (primarily by moisture in the atmosphere) can also affect the energy transport from flames; however work by Frankman et al. (2008) indicates that absorption of thermal radiation from wildland flames by water vapor released in the burning process is less than 16% of the total emitted radiant energy for distances equal to 10-times the flame height. Generally it is accepted that wildland flames thicker than 3 m have an emissivity approaching 0.9–1.0 (Butler et al. 2004; Quintiere 2006; Àgueda et al. 2010).

The physical principals that govern radiant heat transport can also be exploited to quantify these fluxes in the field. Infrared radiometers and thermal imagery are currently the best means to capture the spatial and temporal patterns of radiant heat. Transducers sensitive to different wavelengths can be used to measure radiant fluxes from both flames and surfaces (vegetation, soil) involved in combustion. For example, a radiometer sensitive to the near infrared (between 1 μm and 5 μm) can capture radiant emissions from the combustion products CO_2 (peaks at 2.7 μm and 4.4 μm) and H_2O (peaks at 1.4 μm, 1.9 μm, and 2.7 μm). Other sensors that quantify mid-infrared bands (between 5 μm and 25 μm) are more effective at quantifying the broad-spectrum IR emissions from carbon soot and solid surfaces. However, to accurately quantify these fluxes using IR thermometry requires several considerations for measurements of both flames and surfaces.

First, neither flames nor most surfaces behave as black bodies, and an appropriate calculation of emissivity is required for accuracy. For flame measurements, the estimation of emissivity is complex due to variation driven by fuel chemistry, moisture, and the combustion environment; however, estimates can be made (Parent et al. 2010). Surfaces are easier to characterize and often have an emissivity near 1 (such as 0.95–0.98 for many soils and plant tissues), making accurate measurements more feasible. Even so, IR thermographic cameras and radiometers are subject to the constraints of all optical sensors in that scenes can be obscured by intervening objects or that sensor resolution changes with distance. Distance is a critical consideration as well for another reason: if the field of view includes areas that are not actively burning, radiant heat measurements will be underestimated. Scatter and/or absorption of radiation by the atmosphere can be compensated for, but this procedure can become intractable in the turbulent and smoky environment of wildland fires. Still, capturing radiant energy emission from wildland fires is well developed and a mature sensor technology.

3.2.2 Convective Energy

For laminar flow regimes the energy exchange between a fluid surrounding a solid is primarily conductive, but in turbulent environments the time-averaged energy exchange between hot gases and solid fuels is a special case of conduction that in engineering literature is termed convective heat transfer. Convection is rapid molecule-to-molecule conduction at the fuel surface. The magnitude of energy exchanged through convection is parameterized as being proportional to the product of the temperature difference between the fluid and surface and a scaling factor that is termed the convective heat transfer coefficient: $q = h(T - T_s)$, where q is the convective flux, h is the empirically determined heat transfer coefficient, T is the gas temperature, and T_s is the temperature of the surface. The heat transfer coefficient (h) is typically correlated with flow and fluid properties through relations such as $Nu = hl/k = C\,Re^n\,Pr^m$, where Nu is the Nusselt number, Re is the Reynolds number, a non-dimensional measure of the flow field, n and m are empirically determined exponents, l is a characteristic dimension, and k is the thermal conductivity of the media or fluid. In the case of natural convection from a given temperature source, Re is replaced by Gr, or the Grashof number. $Gr = Bg\,(T_s - T_\infty)\,L^3\,v^{-2}$, where B represents the coefficient of thermal expansion

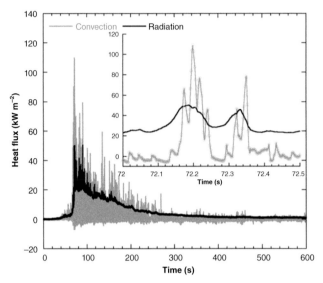

Figure 3.3 Direct measurements of convective and radiant energy from a flame (Frankman et al. 2013).

(approximately $1/T$ for ideal gases), g is the gravitational constant, T_s is the surface temperature, T_∞ is the bulk temperature of the gas, L is the vertical length, and v is the kinematic viscosity. Both Re and Gr are very large in natural convective systems, and often their ratio in some form is used to characterize convective regimes.

Convective heating is a transient phenomenon. Recent work has shown that peak nearly instantaneous convective energy fluxes may significantly exceed radiant fluxes generated from the same flames (Figure 3.3; Frankman et al. 2013). However, as averaging occurs over the fast flickering flame movement, convective heating maxima based on, for instance, 3 s moving averages are nominally 70% of similarly averaged peak radiant heating value.

In a groundbreaking study on the role of convection on wildland fire ignition and spread, Finney et al. (2015) discuss how buoyancy plays a critical part in the ignition process. Based on observations using high speed videography they find that counter-rotating vortex pairs formed as part of the upward plume buoyancy force flames downward into the troughs that are paired with peaks in the flame structure. Canfield et al. (2014) described these structures as analogous to Goertler vortices. These structures form upward- and downward-directed flows that, at least for the downward case, result in rapid convective heating of the unignited fuels ahead of the front. They note that the intermittent heating results in a stepwise upward temperature profile of the unignited fuels immediately ahead of the flame front. This work has generated an entire research direction focused on characterizing the role of convective heating associated with the trough and peak structure in flames.

Capturing convective heat fluxes across time and space in wildland fires remains a challenge and the fluxes remain poorly characterized. While methods for 3D point measurements of convective heat have been well characterized using sensors such as pitot tubes combined with fine wire thermocouples (Butler 2010), capturing extensive convective heat dynamics is still an area of active research. Some promising methods include background-oriented Schlieren photography (Aminfar et al. 2020) and optical flow analyses (Zheng et al. 2019).

3.3 Heating Rates in Flames

While understanding of energy transport in wildland fires remains incomplete (Viskanta 2008), various studies have reported measurements of energy transport from biomass fueled flames. Packham and Pompe (1971) measured radiative heat flux from a fire in Australian forest lands. Heating reached 100 kW m^{-2} when the flame was adjacent to the sensor and 57 kW m^{-2} when the sensor was a distance 7.6 m from the flame (King 1961), no description of flame dimensions were provided. Butler et al. (2004) presented temporally resolved irradiance measurements in a boreal forest crown fire burning primarily in jack pine (*Pinus banksiana*) with an understory of black spruce (*Picea mariana*). Irradiance values reached 290 kW m^{-2}, flames were 25 m tall, and fire spread rates were nominally 1 m s^{-1}. Morandini et al. (2006) measured time-resolved irradiance values from flames burning in 2.5 m tall Mediterranean shrubs (*Olea europea*, *Quercus ilex*, *arbustus unedo*, *Cistus monspeliensis*, and *Cytisus triflorus*). Total and radiative heat fluxes incident on the sensor face were measured but the sensors were not close enough to the flames to experience any convection. Radiative heat fluxes peaked at 7.8, 2.2, and 1 kW m^{-2} for distances to flames of 5, 10, and 15 m, respectively.

Silvani and Morandini (2009) measured time-resolved radiative and total heat fluxes incident on the sensor in four experiments burning pine needles and oak branches. For a burn conducted on a slope of 36% with flame heights of 5.6 m the peak radiative and total heating at the sensor located in the flames were 51 kW m^{-2} and 112 kW m^{-2}, respectively, implying that convective heating was nominally of the order of the radiative heating. Frankman et al. (2012) report measurements from fires burning in a variety of vegetation and terrain. Irradiance beneath two crown fires burning in lodgepole pine (*Pinus contorta*) peaked at 200 and 300 kW m^{-2}, respectively, with flames reaching 30 m; convective fluxes were 15–20% of the peak radiative fluxes. Peak irradiance associated with fires in grasses and leaf and pine needle litter in southern longleaf pine (*Pinus palustris*) reached 100 kW m^{-2}, with a mean value of 70 kW m^{-2} for flames nominally 2 m tall, and convective heating was equal to or greater than the radiative flux. Fires burning in sagebrush-dominated (*Artemisia tridentata* subsp. *wyomingensis*) ecosystems generated peak

Figure 3.4 Time-averaged peak total, radiant, and convective energy from fires in various fuel types (Frankman et al. 2012).
A black and white version of this figure will appear in some formats. For the color version, please refer to the plate section.

radiant energy fluxes of 132 kW m^{-2}, with a mean value of 127 kW m^{-2} for flames less than 3 m tall, and peak convective heating was 20–70% of the radiative heating magnitudes on slopes of 10–30% (summarized in Figure 3.4).

3.4 Fire Models

Many fire spread models have formulated flame models that are best characterized as solid black plates or sheets of uniform temperature and emissivity with specified height, width, and tilt angle. A review of radiant heating models in the context of wildland fire indicates that the solid sheet modeling approach can include significant error (Sullivan et al. 2003). The reasons lie primarily in the assumption of constant temperature and emissivity. In reality radiative energy emission is a volumetric phenomenon; the solid plate approach converts it to a surface phenomenon. It is attractive because it addresses the complex physics associated with definition of the soot volume fraction and particle size as required for volumetric determination of flame irradiance. Additionally, flames are transient, making them difficult to define in geometric terms. Finally, temperature of the flames varies widely throughout the flame volume.

Sullivan et al. (2003) measure peak flame temperatures of 1,583 K in propane flame simulators and explore the effect of flame temperature, emissivity, and location of the surface relative to a tilted flame. Their analysis indicates that the optimum location for the solid radiator model is a vertical surface located midway between the leading edge of the flame base and the flame tip. Flame angle seems to affect radiant energy transport minimally, as do flame widths greater than 20 m. Flame height or length and temperature are difficult to define quantitatively and error in their estimation can be one of the primary sources of uncertainty in modeling radiant energy transport. Others advocate the use of a characteristic flame height equal to one-half the flame length for calculation of radiant energy transport from flames (Gettle and Rice 2002). Perhaps variable temperature and emissivity flame models would be beneficial. Sullivan et al. (2003) conclude that the challenges associated with defining the temporal and spatial fluctuations in these variables precludes any increase in simulation accuracy that would be possible if a model were designed that allowed them to vary with location.

In combustion science, studies of ignition have typically focused on minimum heat fluxes to achieve ignition of a surface. For example, Spearpoint (1999) (see also Quintiere 2006) measured critical ignition threshold heat fluxes of 8–15 kW m^{-2} for various soft and hardwood species with samples nominally 50 mm thick. However, in a unique twist, Cohen performed similar tests on wood shavings exposed to radiant heating alone and found that fine wood shavings exposed to similar and greater heating levels never did ignite. These experiments demonstrated the critical role that natural convection can play in offsetting radiant heating.

3.5 Ignition

From a simplified point of view, the ignition process begins with pyrolysis of the solid biomass. Pyrolysis is thermally driven chemical decomposition that within the context of wildland fire is the conversion of solid material into combustible gases (see Chapter 1). Long chain hydrocarbons comprising the exposed and heated biomass are broken down into smaller molecules through exposure to heat. As the smaller molecules are formed, they eventually accumulate into microscale clouds around the solid material. When the gas cloud is exposed to an ignition source and is of sufficient density the gases react forming a flame. Flaming combustion in natural fuels has been defined as a chemical reaction producing a gas temperatures between 1,500 and 2,500 K in air (Wotton et al. 2012). In the case of wildland fires the reactants (combustible gases formed from pyrolysis) and air are generally separate and only combine at the reaction zone. As a combustible gas density is reached, the oxygen required for the reaction must diffuse to the location of the combustible gases. Because most flow in this environment is turbulent and because the reaction is limited by diffusion of oxygen to the reaction zone rather than chemical kinetics, wildland flames are technically termed turbulent diffusion flames. Flames are most often an ensemble of flamelets with a thickness of approximately 10^{-3} cm, producing energy exothermically at about 10^8 W cm^{-3} (Quintiere 2006).

The chemical reaction can be simply characterized as combustible gases reacting with the oxygen in air to form carbon monoxide, carbon dioxide, and water as well as other trace gases. Generally speaking, nitrogen in the air acts as a heat sink, however in some combustion systems the formation of nitrogen oxide compounds is a critical player in atmospheric pollution chemistry. This is not considered a factor in wildland flames. While it is difficult to characterize biomass from a chemical standpoint, woody material is of the form $C_xH_yO_z$, where x, y, and z are roughly 10, 15, and 7.

Generally speaking, based on these relative quantities of the three major elements, 5–7 kg of air are required for every kilogram of biomass that burns (https://is.muni.cz/el/1423/podzim2013/MEB423/um/Wood_Lesson_02.pdf, last accessed June 10, 2020). When converted to a volume basis this mass of air is equivalent to 5,000–8,000 unit volumes (equivalent volume of fuel burning) of air (at standard pressure and temperature) consumed for every single unit volume of biomass that burns. This has significant implications for understanding wildland fire behavior. The enormous amount of oxygen that is consumed in the reaction means that at every instant more air must be diffusing to the reaction zone to combine with more combustible gases. This ratio of 5,000:1 can provide some insight into the factors that drive wildland fire. Any factor that can increase the availability of oxygen to the reaction zone (say wind) will increase fire intensity. The reverse is also true in that any factor that inhibits the availability of air to the reaction zone will result in reduced intensity and energy release (e.g. dense vegetation, terrain, possibly high elevation). Video footage of large-scale wildland flames clearly indicates the strong interactions between flame intensity and surrounding air flow. Images indicate strong indrafts immediately prior to the arrival of the flame front, followed by a reversal of flow direction (Anderson et al. 2010).

3.6 Flames

Wildland fire flames are unconfined reactions occurring in time and space varying conditions. The boundary conditions are either undefined or varying over time and space. Due to these reasons, energy exchange in wildland flames is dependent on many variables, many of which are dependent on not only the structure (shape, orientation, size, etc.) and chemical physiology, but also the surface properties and burning environment. The impact of energy exchange is also complex, depending not only on static properties but also dynamic changes in the burning material and environment. Simple correlation with variables such as a critical ignition flux must be considered in the context of local material properties and geometry. For example, Douglas-fir wood samples nominally 50 mm in width ignite when exposed to heat fluxes of 12 kW m^{-2}, but smaller samples nominally 1 mm in diameter never ignite, even for exposures exceeding 20 kW m^{-2}. The underlying process is again that the radiant heating incident on the small sample surface is offset by cooling through natural convection. In wildland flames the push and pull between radiant heating and cooling and convective heating and cooling are complex, depending on many environmental and material properties as well as the position, size, intensity, duration, etc., of the flames themselves. In some cases, radiant energy transport will govern the heating, pyrolysis, and ignition processes while, under different fuel and environmental conditions, convective energy transport will dominate these processes.

3.7 Impacts of Heat and Mass Transfer on Flames

Wildland flames are complex, and it is difficult to define boundary conditions as the flames are unconfined. Interaction between energy and mass transport at relatively small scales can translate into macroscale impacts on fire intensity and behavior. Finney and McAllister (2011) summarize the current state of knowledge regarding how fire fronts can interact to form much taller flames, largely due to the reduction of diffusion of air at the center of the merged flame. They discuss the process of merging flames, identifying increases in reaction rate and energy release as separate flames become closer, largely due to enhancement of radiant energy transport between the flames, but once the flames merge the restriction of air to the flaming core results in a stabilization of reaction rates which then dominates the process and results in a steady or decreased burning rate. In any case, the ignition of many spot fires ahead of a spreading fire front will result in increased heat release rates and flame heights, above those levels for individual spot fires.

Another interaction related to mass and energy transport is the generation of fire whirls, as multiple fires draw air into the reaction zone, the inbound air can result in the generation of vorticity that is then conserved and can lead to the formation of fire whirls. When fire whirls form they can result in dramatically increased local energy release at the immediate location of the whirl (Forthofer and Goodrick 2011).

3.8 Conclusion

Characterization of energy transport in wildland flames remains an elusive target. The measurement of energy and mass transport in wildland fires is, at best, tenuous. However, despite the uncertainty associated with the measurements, accessing areas threatened by wildfires or completing prescribed fires, the data that have been reported correlate well with measured consumption of vegetation, they provide upper and lower limits of energy release as a function of vegetation type and burning conditions, and give general information about relative distribution of radiative and convective energy in flames. Of the many relevant research questions associated with wildland fire, one of the most pressing is characterization of the spatial and temporal variability in energy release and how that variability can be represented and replicated in mathematical terms. While sensors have been developed that can sense energy arriving at a point in space and the source of that energy can be associated with nearby flames, it is only with multiarray infrared sensors that there is hope of characterizing spatial energy release. In most cases, those sensors are downward looking and two-dimensional and thus, while the images produced by such sensors are fascinating, there remains the question of what exactly they are sensing in terms of energy release from the vegetation, soil, and atmosphere point of view. Nevertheless, the development of nonintrusive systems shows promise for reduced measurement uncertainty and increased temporal and spatial resolution in laboratory settings. Such systems that can operate successfully in a natural fire environment are currently being developed. Additionally, further efforts to correlate energy release with emissions is an area of increasing concern. With the general public's realization that prescribed burning must be one of the tools for land management, there is a corresponding need to improve tools to predict emissions as a function of burning conditions. So far, all that are available for such correlations are highly general emissions factors that relate emissions production to vegetation consumption. Fire is the source of the emissions and it is only with improved measurements and modeling capabilities to characterize fire parameters, burning conditions, and emissions that successful implementation of prescribed burning at the necessary spatial scales will be achievable.

References

Àgueda, A, Pastor, E, Pérez, Y, Planas, E (2010) Experimental study of the emissivity of flames resulting from the combustion of forest fuels. *International Journal of Thermal Sciences* **49**(3), 543–554.

Albini, FA (1986) Wildland fire spread by radiation: A model including fuel cooling by natural convection. *Combustion Science & Technology* **45**(1–2), 101–113.

Aminfar, A, Cobian-Iñiguez, J, Ghasemian, M, Espitia, NR, Weise, DR, Princevac, M (2020): Using background-oriented Schlieren to visualize convection in a propagating wildland fire. *Combustion Science and Technology* **192**(12), 2259–2279.

Anderson, HE (1969) *Heat Transfer and Fire Spread*. USDA Forest Service, Research Paper INT 69, Intermountain Forest and Range Experiment Station, Ogden UT, 1–20.

Anderson, WR, Catchpole, EA, Butler, BW (2010) Convective heat transfer in fire spread through fine fuel beds. *International Journal of Wildland Fire* **19**, 284–298.

Butler, BW (2010) A portable system for characterizing wildland fire behavior. In: Viegas, DX, ed. *Proceedings of the VI International Conference on Forest Fire Research*; November 15–18; Coimbra, Portugal. Coimbra, Portugal: University of Coimbra.

Butler, BW, Cohen, J, Latham, DJ, Schuette, RD, Sopko, P, Shannon, KS, Jimenez, D, Bradshaw, LS (2004) Measurements of radiant emissive power and temperatures in crown fires. *Canadian Journal of Forest Research* **34**(8), 1577–1587.

Canfield, JM, Linn, RR, Sauer, JA, Finney, M, Forthofer, J (2014) A numerical investigation of the interplay between fireline length, geometry, and rate of spread. *Agricultural and Forest Meteorology*, **189**, 48–59.

Cohen, JD, Finney, MA (2010) An examination of fuel particle heating during fire spread. In: Viegas, DX, ed. *Proceedings of the VI International Conference on Forest Fire Research*; November 15–18; Coimbra, Portugal. Coimbra, Portugal: University of Coimbra.

Finney, MA, Cohen, JD, Forthofer, JM, McAllister, SS, Gollner, MJ, Gorham, DJ, Saito, K, Akafuah, NK, Adam, BA, English, JD (2015) Role of buoyant flame dynamics in wildfire spread. *Proceedings of the National Academy of Sciences* **112**(32), 9833–9838.

Finney, MA, Cohen, JD, Grenfell, IC, Yedinak, KM (2010) An examination of fire spread thresholds in discontinuous fuel beds. *International Journal of Wildland Fire* **19**, 163–170.

Finney, MA, McAllister, SS (2011) A review of fire interactions and mass fires. *Journal of Combustion* **2011**, 548328.

Forthofer, JM, Goodrick, SL (2011) Review of vortices in wildland fire. *Journal of Combustion* **2011**, 14.

Frankman, D, Webb, BW, Butler, BW (2008) Influence of absorption by environmental water vapor on radiation transfer in wildland fires. *Combustion Science & Technology* **180**(3), 509–518.

Frankman, D, Webb, BW, Butler, BW (2010) Time-resolved radiation and convection heat transfer in combusting discontinuous fuel beds. *Combustion Science & Technology* **182**(10), 1–22.

Frankman, D, Webb, BW, Butler, BW, Jimenez, D, Forthofer, JM, Sopko, P, Shannon, KS, Hiers, JK, Ottmar, RD (2012) Measurements of convective and radiative heating in wildland fires. *International Journal of Wildland Fire* **22**, 157–167.

Frankman, D, Webb, BW, Butler, BW, Jimenez, D, Harrington, M (2013) The effect of sampling rate on interpretation of the temporal characteristics of radiative and convective heating in wildland flames. *International Journal of Wildland Fire* **22**, 168–173.

Gettle, G, Rice, CL (2002) Criteria for determining the safe separation between structures and wildlands. In Viegas, DX, ed. *IV International Conference on Forest Fire Research & Wildland Fire Safety Summit*; November 18–23; Luso, Coimbra, Portugal. Rotterdam: Millpress.

King, AR (1961) Compensating radiometer. *British Journal of Applied Physics* **12**, 633.

Morandini, F, Silvani, X, Rossi, L, Santoni, P-A, Simeoni, A, Balbi, J-H, Louis Rossi, J, Marcelli, T (2006) Fire spread experiment across Mediterranean shrub: Influence of wind on flame front properties. *Fire Safety Journal* **41**(3), 229–235.

Packham, D, Pompe, A (1971) Radiation temperatures of forest fires. *Australian Forest Research* **5**, 1–8.

Parent, G, Acem, Z, Lechêne, S, Boulet, P (2010) Measurement of infrared radiation emitted by the flame of a vegetation fire. *International Journal of Thermal Sciences* **49**(3), 555–562.

Quintiere, JG (2006) *Fundamentals of Fire Phenomena*. Hoboken, NJ: John Wiley & Sons.

Raj, PK (2008) Field tests on human tolerance to (LNG) fire radiant heat exposure, and attenuation effects of clothing and other objects. *Journal of Hazardous Materials* **157** (2–3), 247–259.

Silvani, X, Morandini, F (2009) Fire spread experiments in the field: Temperature and heat fluxes measurements. *Fire Safety Journal* **44**(2), 279–285.

Spearpoint, MJ (1999) *Predicting the Ignition and Burning Rate of Wood in the Cone Calorimeter using an Integral Model*. Building and Fire Research Laboratory, NIST GCR 99-775, MD.

Sullivan, AL, Ellis, PF, Knight, IK (2003) A review of radiant heat flux models used in bushfire applications. *International Journal of Wildland Fire* **12**(1), 101–110.

Viskanta, R (2008) Overview of some radiative transfer issues in simulation of unwanted fires. *International Journal of Thermal Sciences* **47**(12), 1563–1570.

Wotton, BM, Gould, JS, McCaw, WL, Cheney, NP, Taylor, SW (2012) Flame temperature and residence time of fires in dry eucalypt forest. *International Journal of Wildland Fire* **21**(3), 270–281.

Yedinak, KM, Cohen, JD, Forthofer, JM, Finney, MA (2010) An examination of flame shape related to convection heat transfer in deep-fuel beds. *International Journal of Wildland Fire* **19**, 171–178.

Yedinak, KM, Forthofer, JM, Cohen, JD, Finney, MA (2006) Analysis of the profile of an open flame from a vertical fuel source. *Forest Ecology and Management* **234**(Suppl. 1), S89.

Zheng, Y, Zhou, X, Ye, K, Liu, H, Cao, B, Huang, Y, Ni, Y, Yang, L (2019) A two-dimension velocity field measurement method for the thermal smoke basing on the optical flow technology. *Flow Measurement and Instrumentation* **70**, 101637.

4

Fire Line Geometry and Pyroconvective Dynamics

JASON J. SHARPLES, JAMES E. HILTON, RACHEL L. BADLAN,
CHRISTOPHER M. THOMAS, AND RICHARD H. D. MCRAE

4.1 Introduction

The shape a wildfire takes as it spreads across a landscape depends on a variety of factors. Chief among these are the weather, topography, and the type, structure, and condition of the fuel, although local variations in, and interactions between, these factors can also have a distinct influence on the evolution of a fire line. The archetypal fire shape is that of a wind-driven fire evolving from a point ignition under uniform conditions. In this case, the fire develops into a roughly elliptical shape, with the major axis of the ellipse aligned with the wind direction and its length-to-breadth ratio determined by the wind speed (Alexander 1985). This shape is commonly exhibited by fires spreading over relatively flat grasslands, for example, but has also been used as a basis to model the evolution of a wildfire's perimeter in other fuel types and under nonuniform conditions. (Anderson et al. 1982; Green 1983; Green et al. 1983; Finney 1998; Tolhurst et al. 2008). Modeling suggests that the elliptical shape arises from wind fluctuations at the edge of the fire causing outward growth, while the fire is stretched in the direction of the wind (Hilton et al. 2015).

Departures from the standard elliptical fire shape can occur for a variety of reasons. For example, inhomogeneities in fuels, variations in topography, and locally modified winds can result in parts of a fire spreading faster or slower relative to other parts. This can lead to the development of concave or convex features in the fire perimeter, which are commonly described as "pockets" or "fingers," respectively (see Figure 4.1). These features can also develop due to fire–atmosphere interactions, as is the case for *convective fingering* (Clark et al. 1996a,

The authors acknowledge the support of the Bushfire and Natural Hazards Cooperative Research Centre, the Australian Research Council and the NCI National Facility at the Australian National University through the National Computational Merit Allocation Scheme supported by the Australian Government. The authors are also grateful to Domingos Viegas for his contribution to aspects of the research discussed in this chapter, and to Kevin Speer for providing comments on earlier drafts of the chapter, which helped improve it.

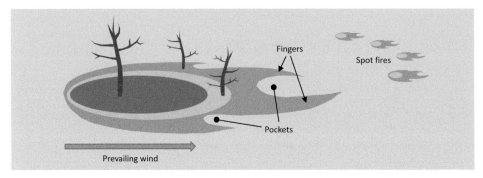

Figure 4.1 Schematic diagram showing the basic shape of a wildfire, along with other features such as pockets, fingers, and spot fires.

1996b). In fact, pyroconvective interactions can have a significant influence on the evolution and shape of the fire perimeter.

Pyroconvection is the buoyant movement of fire-heated air, which is considered as distinct from the ordinary convection that drives atmospheric motions. As the heated air within a fire's plume rises, ambient air is drawn in from the surroundings to replace it. The resulting fire-induced, or *pyrogenic*, circulations comprise the most basic form of fire–atmosphere interaction. These pyroconvective circulations occur in connection with *any* fire – from a burning match to a large-scale wildfire. However, the degree to which pyroconvective circulations can influence the behavior and spread of a wildfire depends on their strength and scale. Of course, the spread of a wildfire is also influenced by the other heat transfer mechanisms: radiative heating of unburnt fuel, conduction of heat within fuels, and advection of heated air by the ambient wind, but the focus of this chapter is on the role of pyroconvection, and in particular how pyroconvective interactions can induce an additional level of dynamism to wildfire propagation – the likes of which are not well-accommodated in current operational fire spread prediction models.

Current operational fire spread models such as Phoenix Rapidfire (Tolhurst et al. 2008), FARSITE (Finney 1998), and Prometheus (Tymstra et al. 2010) are predicated on the assumption that the rate of spread of a wildfire burning in a quasi-equilibrium state can be uniquely determined by the local conditions of fuel, weather, and topography. Indeed, this quasi-steady assumption is central to the development of existing empirical fire behavior models, which establish correlations between fire behavior characteristics and environmental conditions, and which underpin the mechanics of two-dimensional fire simulators. While the quasi-steady assumption can be considered valid (at least for practical intents) over a broad range of fire propagation scenarios, there are now several examples where the assumption can be shown to be invalid. In these instances, the fire does not attain quasi-equilibrium, and the associated spread cannot be accounted for by simple correlations with environmental conditions – we refer to these as instances of *dynamic wildfire behavior.*

Properly accounting for dynamic wildfire behavior in fire-spread prediction is important for several reasons. The occurrence of dynamic fire behavior can result in abrupt increases in the rate of spread of a wildfire, which can catch firefighters off-guard. This can seriously compromise firefighter safety and result in loss of containment of the wildfire (Lahaye et al. 2017, 2018a, 2018b; Page and Butler 2017). Several modes of dynamic fire behavior are subject to environmental thresholds (e.g. relating to wind and topographic slope) and so there is often a very narrow margin between conditions that are favorable for dynamic fire behavior and those that do not support dynamic behavior. As such, a slight change in conditions that are near threshold can result in a significant and dangerous escalation in fire behavior.

Dynamic fire behavior is also closely associated with *extreme fire behavior* – in fact, one could argue that the latter is a subset of the former, that is, instances of extreme fire behavior are instances of dynamic fire behavior (but not necessarily vice versa). This notion appears to be consistent with the thinking of Werth et al. (2011), who defined extreme fire behavior as "Fire spread other than steady surface spread, especially when it involves rapid increases."

Filkov et al. (2020a) went as far as to propose that the term "dynamic fire behavior" be used in place of "extreme fire behavior." Specifically, they defined dynamic fire behavior as a "physical phenomenon of fire behavior that involves rapid changes of fire behavior and occurs under specific conditions, which [have] the potential to be identified, described and modeled" (Filkov et al. 2020a, p. 3). They then stated that "in general, extreme fire (or an extreme fire event) can involve one to several dynamic fire behaviors simultaneously."

Filkov et al. (2020a) further classified dynamic fire behaviors into several specific types, some of which are strongly influenced by fire line geometry and pyroconvective effects:

- Spotting, which is the production of firebrands or embers that are carried by the wind to start new fires beyond the zone of direct ignition by the main fire (NWCG 2012). The rate of production and deposition of firebrands within a wildfire is dependent upon several factors, including winds, fuel dryness, general vegetation type, upper-level weather conditions, and wildfire area (Storey et al. 2020a, 2020b), the last of which relates to the geometry of the burning zone and the pyroconvective power of the fire. Moreover, when spot fires form away from the main fire, they can interact with the main fire and with each other in a manner that is largely driven by pyroconvective influences, dependent upon the particular geometry of the fire lines involved (see Section 4.2.2). This process is known as *spot fire coalescence*. Embers typically follow ballistic trajectories originating in the main fire, but under extreme conditions profuse spotting can become the dominant fire propagation mechanism and embers can be concentrated closer to the ground in a phenomenon that has been termed an *ember storm* (Porterie et al. 2005; Dold et al. 2011). Ember storms present the most dangerous conditions for wildfires to encroach into the wildland–urban interface and initiate structure fires (Chen and McAneney 2004; Gill 2005; Blanchi et al. 2006).

- Fire whirls/tornados, which are spinning vortices of ascending air, flame, and other hot gases and debris characterized by strong vertical vorticity (NWCG 2012; Tohidi et al. 2018). Fire whirls are anchored to a hot fireground and can occur over a range of spatial scales ($\sim 10^{-1} - 10^3$ meters), with large fire whirls able to produce winds of tornadic strength. Fire whirls form in response to interactions between pyroconvective circulations, ambient (horizontal) vorticity, and strong wind shear. Note that large fire tornadoes, such as those reported by Kuwana et al. (2007), Forthofer and Goodrick (2011), McRae et al. (2013), Lareau et al. (2018), and Forthofer (2019), can also involve larger-scale atmospheric convection in addition to strong pyroconvective effects.

- Vorticity-driven lateral spread (VLS, aka fire channeling), which involves rapid lateral spread across a steep leeward slope. Research suggests that the lateral spread occurs in connection with the formation of strong fire whirls, which form due to strong interactions between ambient winds, terrain, and a vigorous fire plume (Simpson et al. 2013, 2014, 2016; Sharples et al. 2015), and that carry the fire across-slope. In addition to the rapid lateral spread, VLS is associated with dense spotting downwind of the lateral spread zone, spot fire coalescence, and the formation of widespread flaming zones (Sharples et al. 2012; Raposo et al. 2015).

- Junction fires, which involve the oblique intersection (or merging) of two fire lines (Viegas et al. 2012; Raposo 2016; Raposo et al. 2018; Filkov et al. 2020a). Research has shown that in some instances the junction zone, where the fire lines intersect, can exhibit very rapid rates of spread due to enhanced radiative heating and pyroconvective interactions (Viegas et al. 2012; Sharples et al. 2013; Hilton et al. 2016). Junction fires also have greater potential to generate fire whirls and enhance spotting (Viegas et al. 2012; Raposo 2016; Pinto et al. 2017; Thomas et al. 2017).

- Eruptive fires, which involve accelerated fire spread up a steep slope or canyon (Viegas and Pita 2004; Viegas 2006). The interaction between the fire plume and steep and/or confined topography can result in plume attachment and a dynamic transition to enhanced and extensive convective heating of fuels upslope from the fire (Wu et al. 2000; Dold and Zinoviev 2009; Sharples et al. 2010; Edgar et al. 2015a, 2015b, 2016; Grumstrup et al. 2017).

- Crown fires, which involve a transition from a surface fire to one that involves the entire vertical profile of a vegetation stand. Specifically, Filkov et al. (2020a) identified active and independent crowning (Van Wagner 1977a) as modes of dynamic fire behavior, and pyroconvective effects are certainly important in determining the likelihood of a surface fire extending into the canopy of a forest. It should be noted, however, that there are a number of quasi-steady fire spread models that can adequately account for the spread of crown fires (e.g. Cruz et al. 2013; Cruz and Alexander 2017), which in our opinion casts some doubts about whether crown fires should be considered as a mode of dynamic fire behavior. As such, crown fires can perhaps be considered as an exception to the rule – where a form of extreme fire behavior does not necessarily qualify as a mode of dynamic fire behavior.

Filkov et al. (2020a) also listed a number of additional dynamic fire behaviors: conflagrations, downbursts, and pyroconvective events. In our opinion these cases require some qualification. For example, a conflagration is the term used to describe a raging and destructive wildfire (AFAC 2012; NWCG 2012) and as such there is nothing that specifies that unsteady fire spread, or rapid changes in fire spread are necessarily part of a conflagration. Hence, in our opinion, while it is certainly likely that dynamic fire behaviors play a part in initiating a conflagration, the conflagration itself should not be considered as a form of dynamic fire behavior.

Similarly, a pyroconvective event – which Filkov et al. (2020a) essentially define as a type of cloud – does not necessarily require dynamic fire behavior. For example, the pyrocumulonimbus event discussed by McCarthy et al. (2018) and McCarthy (2020), which occurred in southern Queensland (Sedgerly Road fire), satisfies the definition of pyroconvective event given by Filkov et al. (2020a), but according to McCarthy (2020) the event was initiated by a number of crown fire runs, which, as discussed, does not necessarily imply the involvement of dynamic fire behavior. Indeed, it seems that the development of the pyrocumulonimbus in this instance was more due to the highly unstable atmospheric environment rather than fire behavioral conditions. This event also stands out as a geographical outlier amongst the catalog of southeast Australian pyrocumulonimbus events (Di Virgilio et al. 2019). The notions of "conflagration" and "pyroconvective event" espoused by Filkov et al. (2020a) appear to be better captured by the definition of "extreme wildfire" provided by Sharples et al. (2016), where an extreme wildfire is one that exhibits widespread active flaming in an atmospheric environment conducive to violent pyroconvection, and which manifests as towering pyrocumulus or pyrocumulonimbus. This involves a coupling between the wildfire and the atmosphere that extends well above the mixed layer and that modifies or maintains the propagation of the wildfire.

We also opine that a downburst should not be considered as a form of dynamic fire behavior as it is essentially a meteorological phenomenon, and one that frequently occurs in the absence of a wildfire. While a downburst over a fire can certainly result in an increase in the rate of spread, the increase is solely due to a sudden increase in wind speed. In essence this situation is no different to a fire responding to any other form of wind gust, and there is nothing to suggest that the fire would respond any differently to what would be predicted by a quasi-steady fire spread model. Also, as pointed out by Kinniburgh (2020), very specific conditions are required for a downburst to impact a fire – especially if the downburst originates in the fire's plume.

The aim of this chapter is to explore these concepts in more detail. We focus on the roles that fire line geometry and fire shape can play in dynamic fire behavior and extreme wildfire development and discuss the specific influences of pyroconvective processes. In the present context, our use of the term "dynamic fire behavior" refers to any instance of fire behavior that does not reflect a state of quasi-equilibrium, and that cannot be adequately described using a quasi-steady modeling approach. We use the term "extreme wildfire" in the sense espoused by Sharples et al. (2016), and acknowledge that an extreme wildfire requires the occurrence of one or more blow-up events, each of which is likely to involve instances of dynamic fire behavior, which result in rapid escalation in fire behavior and rate of spread.

We begin by discussing various modes of dynamic fire propagation, for which pyroconvective interactions are known to play a key driving role. We then extend the discussion to examining the role that dynamic fire propagation plays in the development of extreme wildfires, and how broader-scale convective processes (including pyroconvection) influence the development and subsequent evolution of large coupled fire–atmospheric events (see also Potter 2012a, 2012b). We also provide a review of existing approaches to modeling pyroconvective effects and fire–atmosphere interactions more generally – these include computationally intensive physical and coupled fire–atmosphere models, as well as more recently developed models that favor computational thrift and the possibility of faster-than-real-time operational application. We close the chapter with a discussion of the implications of pyroconvective dynamics for firefighter and community safety, tactical response to wildfires, and strategic wildfire risk management. In particular, a number of avenues for further research are identified.

4.2 Pyroconvective Effects on Fire Propagation

One of the most basic fire propagation scenarios is that of a wind-driven fire line. In this case a line of fire is ignited and allowed to propagate under the influence

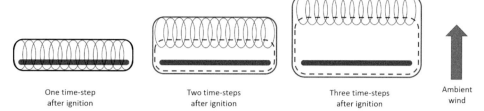

| One time-step | Two time-steps | Three time-steps | Ambient |
| after ignition | after ignition | after ignition | wind |

Figure 4.2 Schematic diagram illustrating the progression of a wind-driven fire line based on the elliptic Huygens principle. The initially straight fire line (thick black line) evolves as the envelope of the elliptic frontlets (some of which are depicted as small ellipses) after each time step. It can be seen that the fire front propagates in a manner that preserves the straight line geometry of the head fire, which is at odds with the parabolic head that is observed to develop in real wind-driven fires that evolve from an initially straight fire line.

of a uniform wind. If the wind direction is roughly perpendicular to the fire line, then the initially straight-line fire will develop into a parabolic front. This change in geometry exhibited by a wind-driven fire line is a key example of how pyroconvective interactions influence the evolution and ultimate shape of a fire perimeter. Indeed, it is worth noting that traditional approaches to modeling fire spread, such as those based on the elliptic Huygens' principle, are not able to account for this basic pattern of fire spread. The reason for this lies in the fact that traditional models assume that different parts of the fire line spread independently, that is, each part of the fire line is assumed to be driven by the local conditions of fuel, weather, and topography, without regard for what other parts of the fire line are doing. This is exemplified in Figure 4.2, which shows how the evolution of a straight wind-driven fire line is handled using the elliptic Huygens' principle. The modeling approach considers each point on the fire line as the source of a "frontlet" that spreads in an elliptical shape, just as if it was an isolated point ignition, with the overall shape of the fire, as time evolves, given by the envelope of all the individual frontlets. Because the driving conditions are uniform across the fire line, each of the elliptic frontlets is identical and, consequently, the modeled head fire remains straight as it evolves.

In reality, however, the straight-line fire develops into a parabolic front due to the fact that different parts of the fire line do not spread independently – rather the fire line evolves as a complex, coupled system. This was first demonstrated using numerical modeling by Clark et al. (1996b), who used a coupled fire–atmosphere model to simulate the evolution of a wind-driven fire. The model coupled a basic fire spread model with the continuity, momentum, and thermodynamic equations that comprise a numerical weather prediction model. Clark et al. (1996b) demonstrated that a pattern of low-level convergence of fire-heated air formed ahead of the center of the fire line. This causes air to be drawn into the fire's plume in such a way that winds are enhanced over the center of the fire line, which then spreads faster in comparison to the edges and provides an explanation for the development of the observed parabolic shape. Moreover, when they considered a sufficiently long fire line, Clark et al. (1996b) found that, instead of developing a single parabolic head, the fire line broke up into multiple parabolic fronts. Clark et al. (1996a) also showed that vertical wind shear can produce rotating updrafts that can act to break up the fire line – in some instances this effect can produce winds of tornadic strength that can rapidly advance the fire front in "dynamic fingers." Although the fire spread model used in these early numerical experiments was rather crude, the fact that it responded to changes in the local wind driven by pyroconvection demonstrated the influence that pyroconvective dynamics can have on the evolution and shape of a fire.

Similar dynamic effects can be observed in windless conditions, for example, when two separate straight-line fires interact with each other. If the two lines of fire are parallel to each other, and are in close enough proximity, then the two fire lines develop curved fronts as they burn toward each other. In this case the convective plumes from each of the fires interact with each other in such a way that pyrogenic winds are induced across the centers of both fire lines.

4.2.1 Junction Fires

If two straight fire lines intersect at an oblique angle, then the interaction between the two fire lines can produce a distinctly dynamic enhancement in the rate of spread in the junction zone. As mentioned in Section 4.1, the combined fire line is referred to as a junction fire, although they are also referred to as "V-shaped fires" owing to their initial shape. Viegas et al. (2012) provided the first experimental demonstration that the rate of spread of the junction point was greater than what would be expected if there was no interaction between the two fire lines. As calculated by Sullivan et al. (2019), if the two straight fire lines do not interact with each other, then the junction point should propagate (inside the V) at a constant rate of spread R given by:

$$R = R_0 \; \mathrm{cosec} \; \theta, \tag{4.1}$$

where R_0 is the basic rate of spread that a single fire line would exhibit under calm conditions and θ is the half-angle between the two intersecting fire lines. Note that Eq. (4.1) represents a purely geometric enhancement of the rate of spread of the junction point compared to the basic rate of spread. Figure 4.3 shows that the values of the (dimensionless, relative to R_0) rate of spread of the junction point measured in the experimental results of Viegas et al. (2012) all sit above the

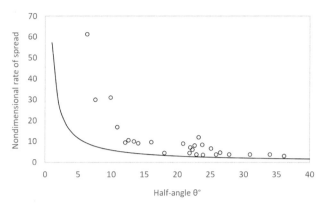

Figure 4.3 Dimensionless rate of spread of the junction point of two intersecting fire lines plotted against the half-angle between the two fire lines. The data (points) is that of Viegas et al. (2012), while the solid line is $\csc\theta$, representing the expected enhancement of the rate of spread due to geometry alone.

$\csc\theta$ line, which implies that there is an additional dynamic enhancement of the rate of spread in addition to the geometric effect.

Viegas et al. (2012) attributed the dynamic enhancement of the rate of spread of the junction point to a concentration of energy (radiation and convection) ahead of the junction zone. Based on this conjecture, they developed an analytical model that provided a reasonably good fit to the observations. However, further work by Raposo et al. (2018) concluded that the very rapid increase in the displacement velocity of the junction point was due to strong convective effects created by the fire, and that radiative effects only became prominent in the later deceleration phase of the junction point as the fire line straightened out.

Thomas et al. (2017) examined large (kilometer-scale) junction fires using the WRF-Fire coupled fire–atmosphere model. In the WRF-Fire model, radiative effects are parametrized within the Rothermel fire spread model (Rothermel 1972) and the interaction between the fire and the atmosphere captured by the model is pyroconvective in nature. Figure 4.4 shows WRF-Fire simulation output for two fire lines that are allowed to interact and two fire lines that are allowed to spread independently, for a case where the initial angle between the two fire lines was $2\theta = 45°$. The results show that the WRF-Fire model predicts a significant enhancement in the rate of propagation of the fire inside the V when the two fire lines interact with one another; after 20 minutes the interacting fire lines had advanced approximately an extra 500 meters beyond the noninteracting fire lines. This supports the notion that the rapid propagation is due to pyroconvective effects. Similar, though less pronounced, results were found when the junction fire was allowed to spread under the influence of an ambient wind (Thomas 2019).

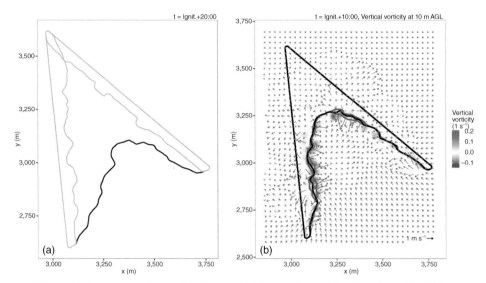

Figure 4.4 Junction fire behavior with $2\theta = 45°$, simulated using WRF-Fire
(Thomas et al. 2017). (a) Grey lines represent the progression of two fire lines
20 minutes after ignition. The evolution of each of the two fire lines has been
simulated separately and in the absence of the other. The black line shows the
position of the simulated fire front at the same time after ignition, that is produced
when the two fire lines are ignited simultaneously, with each allowed to influence
the development of the other. (b) Simulated horizontal wind and vertical vorticity
(10 m AGL) 10 minutes after ignition. The figure shows the formation of counter-
rotating vortex pairs ahead of the fire line, which locally enhance the rate of
forward progression of the fire line.
A black and white version of this figure will appear in some formats. For the color
version, please refer to the plate section.

Considering the more detailed dynamics of the accelerated fire spread within
the V, Thomas et al. (2017) highlighted the influence of small-scale vorticity
couplets – that is, pairs of counter-rotating vertical vortices. These vortex pairs
occurred intermittently at various locations along the fire line but tended to form
ahead of it and were oriented in a way that acted to increase the local rate of
fire spread. Their combined influence acted to rapidly drive the fire forward
(see Figure 4.4). In contrast, analogous simulations involving only a single fire line
(i.e. a single arm of the "V") also developed vorticity couplets, but these formed
behind the fire line and had no effect on the forward advance of the fire line. This
suggests that the geometry of the fire line influences the bulk surface flow in
a way that drives the formation of the vertical vorticity couplets ahead of the fire
line, and that both the bulk surface flow and the emergent vertical vorticity
contribute to the dynamic fire propagation observed in the simulations, and possibly
in real junction fires.

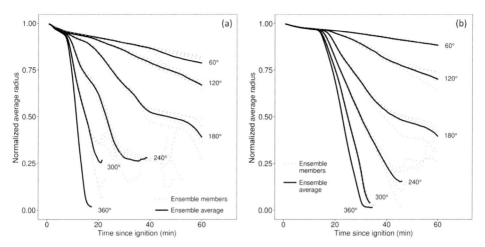

Figure 4.5 Behavior of arc fires simulated using WRF-Fire (Thomas 2019). The figures depict the rate at which the normalized average radius decreases over time, for arc fires of varying angular extents. Dotted lines represent the behavior of individual ensemble simulations and the solid lines the corresponding average for each angular extent. (a) Arc fires with an initial radius of curvature of 500 m, (b) arc fires with an initial radius of curvature of 1,000 m. Steeper curves indicate more rapid fire spread toward the center of curvature, with the 360° cases corresponding to perimeter collapse (see Figure 4.6).

Another type of fire line geometry that exhibits dynamic fire propagation is that of an arc fire. In this configuration, the initial fire line consists of a circular arc of a particular angular extent. For small angular extents the arc fire is similar to a line fire, while an arc fire with an angular extent of 360° is a closed circular "ring" fire, which burns in on itself in a manner that can be referred to as *perimeter collapse*. The coupled fire–atmosphere simulations of Thomas (2019) found that the rate of the inward advance of the fire line (toward the center of the circular arc) increased as the angular extent of the arc increased. This is illustrated in Figure 4.5, which shows the evolution of the inward propagating front of various simulated arc fires. The rate of inward propagation of a circular (360° arc) fire exhibited an approximate 30-fold increase compared to the 60° arc fire. The reason for the differences in behavior appears to be a reduction of entrainment of air into the regions inside the arcs as their angular extent increases and their geometry becomes more closed.

4.2.2 Fire Coalescence

The pyroconvective interactions involved in junction fires, parallel fire line merging and perimeter collapse are all particular cases of those involved in more

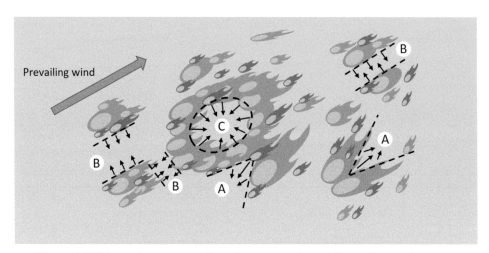

Figure 4.6 Schematic representation of mass-spotting and spot fire coalescence. The dashed lines highlight particular examples of fire line interaction, including: (A) intersecting oblique lines (junction fires); (B) parallel fire line merging; and (C) perimeter collapse (also referred to as a "ring fire").

general fire coalescence, in which multiple fires burning in close proximity combine to form a single contiguous region of flame. This can happen in wildfire situations when a dense ember flux produces numerous spot fires, which then interact and merge in a process called *spot fire coalescence*. This process is illustrated in Figure 4.6. In these situations, mass spotting can act as the dominant fire propagation mechanism and result in a transition from ordinary frontal fire behavior to areal fire propagation, involving large spatial expanses of active flame, and the development of other characteristics of mass fire (Koo et al. 2010, 2012; Finney and McAllister 2011; Storey et al. 2020a, 2020b).

Hilton et al. (2017) considered the effects of pyroconvective interactions on peak fire power in several idealized mass-spotting scenarios. Specifically, they modeled the evolution of the total fire power (i.e. the spatial integral of instantaneous fire line intensity) emanating from different numbers of merging spot fires under two different assumptions: firstly, that there was no pyroconvective interaction between individual spot fires, and secondly with pyroconvective interactions between individual spot fires described using a basic coupled fire–atmosphere model (see Section 4.4.3). Comparison of the results for non-interacting and interacting spot fires indicated that pyroconvective interactions can produce higher peak fire power, and that the effect becomes more pronounced as the number of spot fires increases. This effect is illustrated in Figure 4.7. These results suggest that during mass spotting events the power emitted into the atmosphere by a fire can be significantly greater than what would be expected based on estimates using the bulk rate of spread and Byram's fire line intensity equation (Byram 1959):

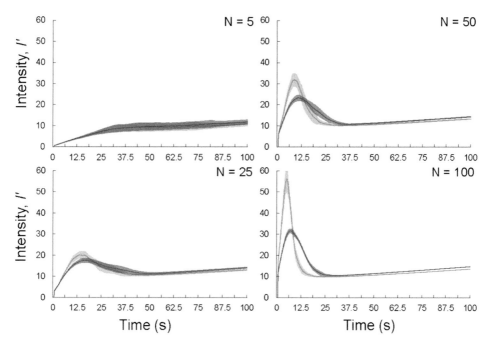

Figure 4.7 Spatial integral of normalized total fire line intensity (fire power) for spot fire coalescence over time for N = 5, 25, 50, and 100 spot fires. The lines show the mean value over 10 ensemble simulations and the shaded confidence bands show one standard deviation. Results from simulations without fire–atmosphere coupling are depicted in dark grey, and results from simulations including fire-atmosphere coupling are depicted in light grey.

A black and white version of this figure will appear in some formats. For the color version, please refer to the plate section.

$$I = HFR,$$

where *H* is the heat of combustion, *F* is the fuel load, and *R* is the rate of spread. This dynamic enhancement of total fire power has implications for the development of extreme wildfires and will be discussed further in Section 4.3. It should be noted that the presence of strong ambient winds will diminish the effect of interaction between multiple spot fires; that is, as the overall fire complex becomes more *wind-driven*.

4.2.3 Fire Whirls

Fire whirls are vertically-oriented, rotating columns of gas and burning debris that form in or near fires (Forthofer and Goodrick 2011). The gases in a fire whirl mainly include air, pyrolysates, and combustion products, the latter of which are often visible as a vortex of flame. Fire whirls occur over a wide range of spatial and temporal scales in the wildfire environment. Small fire whirls, of the order of 1 meter or less in diameter, are a relatively common feature associated with wildland fires – they are often short-lived and, while they can produce local enhancements in rate of spread, they have little effect on the overall propagation of the fire. At the other end of the spectrum, large fire whirls can have diameters of a few hundred meters or kilometers (Goens 1978), produce tornadic wind speeds, and have a considerable effect on the propagation of the fire and dire implications for firefighter and community safety. Indeed, in some instances, very large fire whirls have been characterized as genuine tornadoes; that is, pendant from a cumuliform cloud (Glickman 2000; McRae et al. 2013), while, in other instances, the distinction between a large fire whirl and a genuine tornado is difficult to discern (Lareau et al. 2018; Forthofer 2019).

In addition to the fire, which acts as a pyroconvective fluid sink (see Section 4.4.3), the formation of fire whirls generally requires a source of vorticity (Tohidi et al. 2018). In large wildfires, especially those exhibiting "mass fire" characteristics, there is typically an abundance of vorticity generated by, for example, strong wind shear, frictional shear, and flow around and over topographic obstacles. As this ambient (vertical and horizontal) vorticity interacts with the fire's pyroconvective updraft, it is tilted and stretched (Cunningham et al. 2005; Forthofer and Goodrick 2011; Sharples et al. 2015; Vallis 2017; Tohidi et al. 2018) into intense vertical vorticity. Situations, or locations, particularly prone to fire whirl formation include: the leeward side of prominent ridges (Countryman 1971; Simpson et al. 2013), interaction of multiple fires (Liu et al. 2007), and in the wake of a topographic obstacle or large plume (Tohidi et al. 2018). Phenomena similar to fire whirls have been observed in the closely related system of a jet of gas entering a moving stream of air (Cunningham et al. 2005), suggesting that some whirls may be purely a product of the presence of a plume of hot air rapidly rising in a cross-wind.

Fire line geometry can also have a strong effect on the potential for fire whirl formation. Tohidi et al. (2018) specifically note the prevalence of fire whirls that form in association with an "L-shaped" fire (or "V-shaped" fires, more generally). They identify three types of fire whirl: Type I, which form at the ends of the arms of the L; Type II, which are unsteady whirls that travel along the edges or flanks of the fire and are intermittently shed, sometimes persisting for some distance away from the fire; and Type III, which are stable whirls that form away from the main fire, typically between the two arms of the L. Other fire whirls can be characterized as one of these three types – for example, Type II fire whirls often form intermittently in the wake of a significant fire plume, while Type I fire whirls can form in response to the interaction between multiple coalescing spot fires (see figure 3 of Tohidi et al. 2018).

Fire whirls present a significant threat to firefighter safety. Depending on their size, they can produce wind speeds of less than 10 ms^{-1} to 50 ms^{-1} or more. The pyrotornado documented by McRae et al. (2013) generated wind speeds estimated at EF3 (tornado) strength, while Lareau et al. (2018) reported that the large whirl associated with the 2018 Carr fire generated winds in excess of 64 ms^{-1}, which again categorizes the event as equivalent to an EF3 tornado. This fire whirl caused several fatalities and injuries, as did a large fire whirl reported in association with the Green Valley fire, which burnt during the devastating 2019/20 season in southeast Australia and is reported to have flipped a 15-tonne firefighting tanker.

Large fire whirls can also carry burning debris, including large branches, and deposit them some distance away from the main fire, thereby exacerbating the prevalence of spot fires. This can lead to a rapid escalation in the size and intensity of the fire, which can then compromise the safety of nearby firefighting crews, as well as the broader community, and increase the threat to the wildland–urban interface. Indeed, some structures within the suburb impacted by the 2003 Canberra pyrotornado (McRae et al. 2013) exhibited significant wind damage alone, while others were damaged by a combination of fire and wind. In this respect, it is worth noting that the current Australian Standard for building houses in bushfire prone areas does not account for the effect of strong pyrogenic winds (Sharples and Cechet 2017).

4.2.4 Eruptive Fires

Pyroconvective processes can also influence fire propagation through topographic interactions, and the first example of this involves fire propagation up steep and confined slopes. Under typical circumstances, as a fire spreads up a slope the convective plume rises into the air and the fire spread is driven by radiative transfer and by relatively short-range convective transfer of heat into unburnt fuels. For slopes up to about 20–25°, the effects of radiative and convective heating on fire rate of spread increases as the slope increases due to the closer proximity of the plume to the terrain surface. This effect has been described using empirically derived formulas (e.g. Van Wagner 1977b) and rules of thumb such as that forwarded by McArthur (1967), which posits that the rate of spread approximately doubles for each 10° increase in slope.

However, it has been shown that for slopes greater than about 25° entrainment imbalances can result in a distinctly different mode of fire propagation (Wu et al. 2000; Grumstrup et al. 2017). In basic terms, once the slope exceeds a certain threshold, the downslope entrainment of air into the fire's convective plume cannot be matched by the upslope entrainment, and as a consequence the fire's plume, and the flames within it, attach to the terrain surface in a process known as *plume attachment* or *flame attachment*. The effect can be more pronounced in confined topography such as canyons or trenches as the sidewalls of the canyon can inhibit lateral air entrainment into the plume, exacerbating the imbalance between upslope and downslope entrainment into the plume.

Research indicates that the transition to plume attachment is highly sensitive to slope, with experiments and numerical simulations indicating a threshold inclination of 24–26° (Wu et al. 2000; Grumstrup et al. 2017; Edgar et al. 2015a, 2015b, 2016). Hence at a slope of about 20° a fire's plume can be separated from the surface and rise directly into the air, while on a slope only a few degrees steeper the plume will be attached. Once the plume attaches to the surface, fire propagation is dominated by convective heat transfer – the fuels upslope of the fire are bathed in the hot plume of the fire for a considerable distance ahead of the fire. As these fuels rapidly ignite they can reinforce the attachment effect resulting in a positive feedback that can cause the fire to accelerate considerably; indeed, Dold and Zinoviev (2009) were not able to assign an upper limit for the rate of spread a fire might attain under such circumstances.

The abrupt acceleration that can accompany plume/flame attachment and associated high rates of spread have been implicated in several critical firefighter safety incidents (Viegas et al. 2009; Sharples et al. 2010; Page and Butler 2017). These instances of rapid increases in rate of spread associated with steep slopes and canyons have become known as *eruptive fire* events, or simply as *fire eruption*. Research is still ongoing to better understand how the threshold angle of inclination for attachment is influenced by terrain geometry; for example, in rectangular trenches attachment takes place at angles at 26° and above (Drysdale et al. 1992; Simcox et al. 1992; Edgar et al. 2015a, 2015b, 2016), while experiments in V-shaped canyons suggest flame attachment occurs at a higher threshold angle (see Figure 4.8).

The presence of wind can also affect the threshold angle of attachment for eruptive fire behavior. However, at this point there is no definitive research that addresses this problem in a systematic way. In the absence of a definitive study it is perhaps sufficient to consider existing research that relates flame angle to the prevailing wind speed and the intensity of the fire (e.g. Albini 1981; Martin et al. 1991; Weise and Biging 1996). The flame angle can then be subtracted from the threshold angle of attachment for calm conditions to provide an estimate for the threshold attachment angle in the presence of wind. At the very least, this could be used as a starting point for further scientific inquiry.

4.2.5 *Vorticity-Driven Lateral Spread*

Vorticity-driven lateral spread (VLS) involves rapid lateral fire propagation across steep, leeward-facing slopes (Sharples et al. 2012; Simpson et al. 2013, 2014,

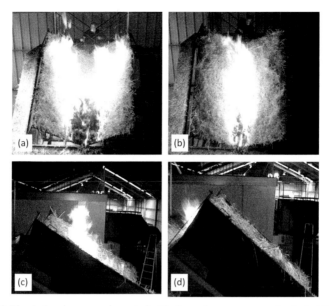

Figure 4.8 Fire behavior experiments in V-shaped canyons conducted at the Center for the Study of Forest Fires in Lousã, Portugal (University of Coimbra). Ignition was a point ignition in the centerline at the bottom of the canyon. (a) Plan view of fire spreading up a canyon inclined at 30° showing two distinct head fires; (b) Plan view of fire spread up a canyon inclined at 40°, showing a single head fire spreading up the canyon centerline; (c) Side view of the fire in the 30° canyon showing the flames rising vertically; (d) Side view of the fire in the 40° canyon showing the flames attached to the surface and erupting out of the top of the canyon.
A black and white version of this figure will appear in some formats. For the color version, please refer to the plate section.

2016), which has the effect of broadening the lateral expanse of a wildfire. In addition, ember production can be enhanced as burning pieces of bark, leaves, and small branches are torn asunder by the strong turbulence associated with VLS. This burning debris can then be transported by the turbulent flows associated with VLS and the prevailing winds, resulting in mass spotting downwind from the lateral spread zone. The dense spot fires so formed then interact, coalesce, and form deep flaming zones (see Figure 4.9).

The lateral spread associated with VLS arises due to a three-way interaction between the wind, the terrain, and an active fire. As sufficiently strong winds flow over rugged terrain, the flow can separate over leeward slopes forming separation eddies and producing ambient horizontal vorticity in the lee of ridge-tops. Pyroconvective interactions between the fire's plume and this horizontal vorticity can then produce vertical vorticity, through vortex tilting, and an increase in the strength of the vorticity through vortex stretching (Vallis 2017; Tohidi et al. 2018).

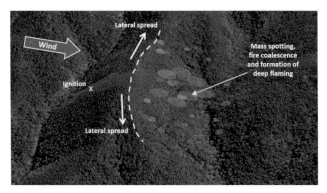

Figure 4.9 Schematic diagram illustrating a typical scenario involving VLS and associated downwind spotting, spot-fire coalescence, and formation of deep flaming. The fire has ignited mid-slope on a windward facing slope and initially spreads up the slope with the wind. As the fire encounters the ridge line (white dashed line), dynamic fire-atmosphere interactions drive the fire laterally (VLS). The regions of lateral spread act as an enhanced source of embers, which are deposited downwind as a dense ember attack.
A black and white version of this figure will appear in some formats. For the color version, please refer to the plate section.

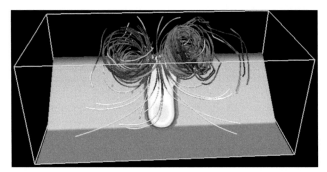

Figure 4.10 Coupled fire–atmosphere model (WRF-Fire) output with wind tendrils showing a counter-rotating pair of vortices at the ridge crest of a leeward slope. The fire has been ignited at the bottom of the leeward slope and has burnt back up the slope against the prevailing winds. The counter-rotating vortices are carrying the fire laterally across the top of the leeward slope. The vortex on the right is spinning in a clockwise direction (i.e. with negative vertical vorticity), while the one on the left is spinning in an anti-clockwise direction (i.e. with positive vertical vorticity).
A black and white version of this figure will appear in some formats. For the color version, please refer to the plate section.

Typically, this results in the formation of one or two vortices, essentially fire whirls, that can intermittently travel across or down leeward-facing slopes (Countryman 1971) carrying flames and embers with them. Figure 4.10 shows the formation of a counter-rotating pair of vortices in response to a fire burning up a leeward slope in a coupled fire–atmosphere model. At the point where the fire meets the ridgeline, strong pyroconvective interactions between the fire's plume and the vorticity caused by flow separation (lee slope eddies) produce the vertical vorticity that acts to spread the fire laterally (i.e. at right-angles to the wind direction) across the slope.

The intermittent nature of lee slope eddies means that VLS is driven by strong but intermittent winds, with instances of higher wind speeds contributing to greater relative shear, enhanced turbulence, and more intense fire behavior. Hence, in addition to spreading the fire laterally across the leeward slope and increasing ember production, VLS also allows embers to be lofted higher into the air – indeed, a dark cumuliform smoke column, indicative of high fire intensity and violent pyroconvection, is a distinguishing feature of VLS (Sharples et al. 2012). The combined influence of all of these factors results in an increase in the width of the fire from which firebrands are generated, the number that are produced, and the distance they can be transported within the plume and by the prevailing winds. Hence, there is a strong connection between VLS and mass spotting, which then results in expansive areas of active flaming and significant increases in the spatial integral of instantaneous energy release; that is, the *power* of the fire.

4.3 Extreme Wildfire Development

Extreme wildfires are distinguished by their disproportionate impacts on environment and society. Although relatively rare, extreme wildfires are consistently associated with the highest levels of economic and property loss, damage to natural resources and human infrastructure, and fatalities – including human, livestock, and wildlife losses (Cruz et al. 2012; Williams 2013; Tedim et al. 2018). Extreme wildfires also play an important part in the evolution of mega-fires (Attiwill and Adams 2013; Attiwill and Binkley 2013; Williams 2013), even though they may only manifest over short episodes within the weekly or monthly timescales associated with large campaign fires. Sharples et al. (2016) defined an extreme wildfire as one that exhibits widespread active flaming in an atmospheric environment conducive to the development of large-scale, violent pyroconvection. This definition reflects the balance between the atmospheric conditions – primarily atmospheric instability and moisture – and the spatial wildfire dynamics that are required for the development of towering pyrocumulus (pyroCu) or pyrocumulonimbus (pyroCb).

It is of interest to note that the World Meteorological Organization (WMO) disfavors use of the terms "pyrocumulus" and "pyrocumulonimbus" (in part due to the mix of Greek and Latin etymology). The official WMO terminology for a towering pyrocumulus is *"cumulus congestus flammagenitus,"* while a pyrocumulonimbus has the official term *"cumulonimbus flammagenitus."* Pyrocumulo-

Figure 4.11 Pyrocumulonimbus with anvil (*cumulonimbus incus flammagenitus*) over the 2013 Grampians fire, Victoria. Photo: Randall Bacon
A black and white version of this figure will appear in some formats. For the color version, please refer to the plate section.

nimbus clouds can then be further described as "*cumulonimbus calvus flammagenitus*" – a fire plume that has developed into a cloud capable of producing precipitation and lightning but that has not yet reached the tropopause, and "*cumulonimbus incus flammagenitus*" – one that has reached the tropopause and developed an anvil cloud (see Figure 4.11). A pyrocumulonimbus that develops an anvil will also often exhibit an overshooting top, which can inject smoke and other pyrogenic particulates into the stratosphere (Fromm et al. 2005, 2006, 2008a, 2008b, 2010).

Extreme wildfires involve a complex coupling between the fire at the surface and the free troposphere, which directly contributes to the extreme fire behavior exhibited by these fires and their transformative impacts on environment and society. These fires do not adhere to the traditional quasi-steady conceptualization of fire spread and may not even possess a well-defined fire front. Indeed, extreme wildfires tend to propagate as a cascading series of medium- to long-range spot fires with no clear delineation between burnt, burning, and unburnt fuels. In these cases, the extent of a wildfire is best described as the envelope of all of the burning regions, sometimes referred to as a "pseudofront" (Luke and McArthur 1978), recognizing that it may evolve rapidly in a discontiguous manner. As such, extreme wildfires can produce unexpectedly severe fire behavior and require new tools to forecast their occurrence, how they spread, the threats they pose, and how to mitigate their impacts.

4.3.1 Role of Wildfire Dynamics

Pyroconvective processes are critical in the development of extreme wildfires. At the scale of the wildfire (10s–100s of meters), pyroconvective interactions between

Figure 4.12 Left: The 1994 South Canyon fire in Colorado (Butler 1998) – an example of a blow-up fire (Photo: SCFAIT (1994)). Right: The 2013 Forcett-Dunalley fire (Ndalila et al. 2020) – an example of an extreme wildfire. Photo: Rebecca White
A black and white version of this figure will appear in some formats. For the color version, please refer to the plate section.

the fire and the environment and between different parts of the fire itself can lead to rapid escalations in fire behavior known as "blow-up events." The standard definition (AFAC 2012; NWCG 2012) notes that blow-up events are *often accompanied by violent (pyro)convection and may have other characteristics of a fire storm*, although it should be noted that the violent pyroconvection associated with a single blow-up event will typically be of a smaller scale compared to that associated with an extreme wildfire. Figure 4.12 provides a visual comparison of a blow-up fire event with an extreme wildfire event. In the blow-up event, the photographer's vantage point is only a few hundred meters away from the fire, while in the extreme wildfire event the photographer is tens of kilometers away from the fire. The dimensions of the plume emanating from the singular blow-up event are only of the order of 0.25 km^3 (assuming dimensions of about 500 m $\times$ 500 m $\times$ 1,000 m), while pyrocumulonimbus plumes can be of the order of 100s–1,000s km^3, which comprises a difference of about three or four orders of magnitude.

In terms of local wildfire dynamics, blow-up events driven by pyroconvective interactions can trigger a transition from ordinary frontal fire behavior to a more "areal" pattern of burning, where active flaming occurs quasi-simultaneously over large tracts of the landscape (up to several km^2). These instances have been referred to as mass fires (Finney and McAllister 2011) in the literature, where the term applies to both wildfires and large urban conflagrations. McRae and Sharples (2013, 2014) and McRae et al. (2015) used the term *deep flaming* to refer to this phenomenon and highlighted its role in the development of several extreme wildfire events. Deep flaming events result in large spatial integrals of instantaneous energy release from the surface, which have implications for the way these fires interact with the atmosphere.

A number of fire behavioral processes that can trigger deep flaming events have been identified, many of which involve modes of dynamic fire propagation, as discussed in Section 4.2. These triggers include (McRae and Sharples 2013, 2014; Badlan et al. 2017, 2019, 2021a, 2021b):

- Very strong winds: assuming the flaming residence time is not affected, strong winds have the potential to advance the head fire more rapidly than the back of the flaming zone, resulting in an increase in the area of the flaming zone. However, the strong winds required for this to occur may inhibit the vertical development of the plume – hence the fire may still propagate as a wind-driven fire (perhaps with profuse spotting) rather than developing into an extreme wildfire in the strict sense defined by Sharples et al. (2016).
- Change in wind direction: a significant change in wind direction, such as occurs when a cold front impacts a fire, can transform the flank of a fire into an extensive, fast-moving head fire. This scenario is common in southeastern Australia (particularly Victoria) where hot north-westerly winds drive wildfires toward the southeast ahead of a cold front. When the cold front arrives, the north-westerly wind direction changes to a south-westerly, and the long north-eastern flank, which can be tens of kilometers in extent, is transformed into a head fire. The long fire line means that the fire can spread at a very high rate (Cheney et al. 2001; Canfield et al. 2014) and produce a rapid increase in the area of the flaming zone. Frontal changes can also result in an increase in atmospheric instability, enhancing the likelihood of extreme wildfire development.
- Eruptive fire behavior: the sudden terrain-driven acceleration in the fire's advance up a steep slope or canyon can have a similar effect on the area of the flaming zone as that produced by very strong winds.
- Mass spotting and fire coalescence: numerous spot fires ahead of the main fire zone, produced by massive ember fluxes (with possible contributions from lightning), can interact and coalesce in a way that enhances the total power output from the fire (Hilton et al. 2017) and produces expansive areas of active flaming (Finney and McAllister 2011). Profuse spotting can act to trigger blow-up events and extreme wildfire development and can also be maintained or enhanced through pyroconvective processes in the active phase of an extreme wildfire. This is a key example of how an extreme wildfire can maintain its own propagation as an extreme wildfire.
- Vorticity-driven Lateral Spread (VLS): the rapid lateral spread of the fire across lee-facing slopes is often accompanied by dense spotting downwind of the lateral spread zone. This can be attributed to enhanced ember generation driven by

strong vorticity in the lateral spread zone, increased fire intensity producing stronger pyroconvective updrafts, which launch and carry embers aloft, and embers being incorporated into the prevailing winds and deposited downstream. Mass spotting and fire coalescence can then result in expansive zones of active flaming. (Sharples et al. 2012; Simpson et al. 2013, 2014, 2016).

- Inappropriate or overzealous use of incendiaries: during the application of prescribed fire, either as part of strategic hazard reduction or tactical backburning, inappropriate use of incendiaries can result in the development of deep flaming. This can occur for example if spot ignition patterns are too dense in hazard reduction operations, or if long lines of fire are introduced too rapidly during backburning under dangerous conditions. Fire line geometry can also be a factor, for example when junction fires are produced. Inappropriate or overzealous use of incendiaries can exacerbate other triggers of deep flaming such as mass spotting and fire coalescence. Indeed, unwitting application of incendiaries (e.g. as part of backburning operations) in regions prone to VLS can have particularly dire consequences.

McRae et al. (2015) demonstrated a spatiotemporal link between instances of deep flaming, inferred from large areas of uniform spectral signature in multispectral line scan data, and the development of extreme wildfires. They showed that the times when the maximum plume height approached the tropopause coincided with times when different parts of the fire complex were exhibiting deep flaming. Moreover, radar data indicated that the highest radar return intensities were located immediately downwind of the regions of deep flaming. In this study, three instances of deep flaming were identified: two were connected with VLS occurrence and one with a mass spotting event – the two instances associated with VLS generated pyrocumulonimbus.

The link between deep flaming and extreme wildfire development was also examined by Badlan et al. (2017, 2019, 2021a, 2021b). Their work considered the development of convective plumes in a numerical weather prediction model in response to artificial surface heat fluxes of different intensities and geometric configurations. Badlan et al. (2017, 2021a) found that, when the intensity was held constant, the plume emanating from a heat flux in a solid circular configuration was more likely to penetrate higher into the troposphere as the radius of the circle increased. This is not unexpected, as maintaining the same intensity over a larger area increases the total fire power and increases the strengths of the updrafts within the plume, thus allowing the plume to rise to a greater altitude. Badlan et al. (2019, 2021b) further examined the combined influence of fire power and fire shape by modeling the effect on the plume when the total fire power was held constant, but the geometric configuration of the surface heat flux was varied. Heat fluxes were configured over various rectangular domains of equal area but with different perimeter-to-area ratios described by the isoperimetric ratio,[1] with a larger isoperimetric ratio more representative of fires exhibiting deep areal flaming, and a smaller isoperimetric ratio more akin to ordinary lineal fire fronts. Badlan et al. (2019, 2021b) found a negative log-linear relationship between maximum plume height and isoperimetric ratio, which indicates that the plume emanating from a fire exhibiting deep flaming is more likely to reach higher into the atmosphere than one emanating from a typical fire front, even if the total power of each of the fires is the same. This suggests that deep flaming is a critical component of extreme wildfire development.

Plumes that rise higher into the atmosphere are more likely to be affected by secondary convective processes such as those driven by latent heat release. Fire plumes contain a significant amount of water vapor, which originates from evaporated fuel moisture, moisture created as a by-product of combustion, and moisture entrained into the plume from the surrounding air (Potter 2005; Luderer et al. 2006; Cunningham and Reeder 2009). As a plume rises into the atmosphere, it expands and cools, and if it rises high enough the water vapor in the plume will condense into liquid water droplets. At this point, pyrocumulus clouds form within the plume and the release of latent heat enhances the buoyant acceleration of air parcels within the plume, causing it to rise further into the atmosphere. As the plume rises higher it continues to expand and cool, leading to further condensation and latent heat release, which, if conditions are conducive, can cause the plume to develop into towering pyrocumulus or pyrocumulonimbus. Pyrocumulonimbus formation requires the plume to rise very high into the upper levels of the troposphere, where the formation of ice particles leads to electrification of the plume. At this stage the plume has developed into a thunderstorm and can exhibit many of the usual characteristics of thunderstorms – strong, blustery winds, lightning and hail, in addition to characteristics unique to fire-generated thunderstorms such as dense ember fluxes and highly erratic fire behavior.

The level at which pyrocumulus initiation takes place was considered by Lareau and Clements (2016), who used direct observations to determine that the convective condensation level (CCL), rather than the ambient lifted condensation level (LCL), was a better estimate of the initiation height. They also noted that factors such as variations in humidity, ambient wind shear, and the environmental temperature and moisture profiles can all exert significant influence on the onset and development pf pyroconvective clouds, just as they do on other forms of deep,

---

[1] The isoperimetric ratio of a simple closed curve is L^2/A, where L is the perimeter length and A is the area enclosed by the curve. This ratio is a dimensionless constant and is invariant under scaling of the size of the curve.

moist convection. In the most extreme cases, pyroconvective plumes can penetrate into the lower stratosphere before they lose buoyancy. This stratospheric penetration can result in the injection of smoke and other pyrogenic particulates into the stratosphere where they can have broader scale effects on climatic processes (Fromm et al. 2006; Luderer et al. 2006; Trentmann et al. 2006).

Finney and McAllister (2011) noted a number of specific effects of fire interaction in mass fires. Of particular relevance here is the influence that fire interactions can have on burning and heat release rates, flame dimensions, and indraft velocities. Experiments using stationary fires (e.g. wood cribs) have shown that fire–fire interactions can dramatically alter the rate at which fuels burn. While these experiments are not explicitly representative of wildland fires, the heat transfer mechanisms and the interactions they drive can be considered as similar to those present in wildland fires, and so the results strongly suggest that similar effects can be expected to occur in wildfire situations. Finney and McAllister (2011) describe several studies where burning rate and heat-release rate were shown to increase as the separation between individual fires decreased – in some cases, burning rates increased by 400% when the separation between individual fires was halved. Even as individual fires merged completely, the burning rate was still observed to be higher than if the fires were burning independently without any interaction. Increases in the rate of heat release from a fire can cause the plume to reach higher into the atmosphere where secondary convective processes can promote the development of extreme wildfires – this is especially the case if burning rates are enhanced over large spatial expanses. Liu et al. (2007) attribute the increases in burning and heat-release rates to the dynamic enhancement of radiative and convective heating, although the development of strong pyrogenic vorticity can also act to enhance burning rates. Enhanced heat release and the effects of pyrogenic vorticity can then increase flame lengths and the strength of the indraft into the fire's plume (Finney and McAllister 2011). These effects have further implications for the development of extreme wildfires.

The apparent connection between deep flaming and extreme wildfire development, which comprises one or more blow-up fire events, was used as the basis for the development of the conceptual model of McRae and Sharples (2013, 2014). The Blow-Up Fire Outlook (BUFO) model takes the form of the binary flowchart in Figure 4.13. The BUFO model starts with the prerequisite condition of an uncontrolled fire burning on a day of elevated fire danger. In terms of the current Forest Fire Danger Index (FFDI) used in southeast Australia, elevated fire danger is taken as FFDI >25, indicating conditions under which fire control would be difficult. The model then assesses the likelihood of deep flaming occurrence by checking the conditions relating to the known triggers discussed in the above bullet list. The presence of rugged terrain, which is defined

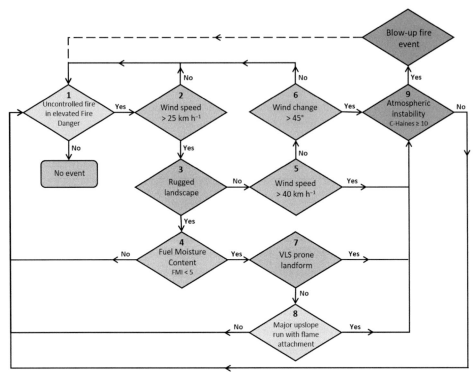

Figure 4.13 The Blow-Up Fire Outlook (BUFO) process model. The model is intended to be used on an iterative basis (e.g. every hour) to assess the likelihood of blow-up in an active wildfire. Implementation starts at node 1 and proceeds through subsequent nodes in the manner of a binary decision tree.

as terrain with high local relief, is checked because it is conducive to mass spotting and increases the likelihood of dynamic fire behaviors like eruptive fire behavior and VLS. Fuel moisture content is also assessed since critically low fuel moisture content enhances ember viability; that is, the likelihood that embers develop into spot fires when they land in unburnt fuel. Critically low fuel moisture content here is defined in terms of the fuel moisture index (FMI) (Sharples et al. 2009; Sharples and McRae 2011), with FMI ≤ 5 taken as the threshold value (see Figure 4.13, node 4). The last component of the BUFO model assesses the condition of the atmosphere to see if it is conducive to deep moist convection. In the current form of the BUFO model, the condition of the atmosphere is gauged using the continuous Haines (c-Haines) index (Mills and McCaw 2010). The BUFO model uses a threshold c-Haines value of 10, which is taken as a reasonable approximation to the 95th percentile of values over southeastern Australia. Mills and McCaw (2010) showed that extreme c-Haines values were associated with a larger proportion of pyroCu fire events, while Di Virgilio et al. (2019) showed that pyroCb events did not occur when c-Haines values were less than about 8, and were more likely for c-Haines >10.

The performance of the BUFO model was assessed in an operational trial conducted by the Australian Capital Territory Emergency Services Agency and the New South Wales Rural Fire Service informally between December 2013 and March 2016. Over this period, 23 assessments were conducted, resulting in three alerts being issued. The fires covered by two of these went on to develop blow-up fire events. The Probability of Detection was 100%, the False Alarm Rate was 33%, and the True Negative Rate was 100%. At the same time, a set of pyroCb events outside the trial were successfully assessed. Agencies involved in the operational trial now use the model operationally to augment core predictive systems. The BUFO model was also used to provide a retrospective account of the conditions surrounding the 2017 Pedrogão Grande fire in Portugal (Boer et al. 2017; Turco et al. 2019) and was found to successfully predict the blow-up that resulted in 66 fatalities (P. M. Fernandes, Personal Communication). The BUFO model is being revised to include lessons from the unprecedented events of the 2019–2020 Australian bushfire season.

Atmospheric conditions conducive to extreme wildfire development, and pyroCbs in particular, have also been considered by Peterson et al. (2017), Tory (2019), and Tory and Kepert (2021). Peterson et al. (2017) studied the occurrence of intense pyrocumulonimbus activity over western North America and introduced a meteorology-based conceptual model for pyroCb development. Particular features identified as conducive to pyroCb development include extreme surface fire danger (hot, dry, and windy) and the presence of an intense plume-dominated wildfire burning in a deep mixed layer with minimal wind shear in the lower troposphere and a lower-tropospheric lapse rate close to dry adiabatic. Mid-level and upper-tropospheric features required for intense pyroCb development include mid-level moisture, high potential instability, and divergence in the upper-troposphere.

Peterson et al. (2017) note that the mid- and upper-tropospheric conditions resemble those associated with the development of traditional dry (non-pyroCb) thunderstorms, but that the increased thermal buoyancy from large and intense wildfires can act as a trigger for pyroCb formation even in the absence of traditional triggering mechanisms. So, while the conceptual model of Peterson et al. (2017) doesn't explicitly consider the likelihood of wildfire dynamics conducive to deep flaming per se, it does acknowledge the necessity of a large and intense wildfire to trigger intense pyroCb development. The findings of Di Virgilio et al. (2019) also suggest that extreme surface fire danger may not be necessary for pyroCb development. Their studies of pyroCbs over southeastern Australia indicated that pyroCbs could develop under surface fire danger that was classified as only High to Very High.

Tory (2019) and Tory and Kepert (2021) developed an index to assist in prediction of pyroCb development, which was based on the well-known Briggs equation (Briggs 1975) for describing the centerline of a convective plume:

$$z_c = \frac{1}{U} \left[\left(\frac{3}{2\beta^2} \right) \frac{H_{flux}}{\pi \rho c_p \theta_a} \right]^{\frac{1}{3}} x^{\frac{2}{3}},$$

where z_c is the height of the plume centerline a distance x away from the convective source, H_{flux} is the heat flux (fire power) produced by the fire, ρ, c_p, and θ_a are the density, specific heat and potential temperature of the ambient air, U is the horizontal wind speed, and β is an entrainment parameter. The free convection height z_{fc} can be related to the plume center height and the Briggs equation can be rearranged to obtain an expression for the minimum (threshold) fire power required for the plume to reach the free convection height, beyond which the plume is able to rise of its own accord and develop into towering pyroCu and pyroCb. The threshold fire power is (Tory 2019; Tory and Kepert 2021)

$$P^* = C \left(z_{fc} \right)^2 U \, \Delta\theta,$$

where C is a constant, and $\Delta\theta$ is the difference between the average potential temperature of the plume at the free convection height and the potential temperature of the ambient air. The Briggs plume equation was used in a similar way by Lareau and Clements (2017) to relate observed plume height to an equivalent fire-generated sensible heat flux.

Implementation of P^* also incorporates a secondary index based on the dry eucalypt fire behavior model of Cheney et al. (2012). The purpose of this secondary index is to filter out surface weather conditions that produce low values of the calculated fire power threshold, but for which uncontrolled wildfire propagation would be extremely unlikely (Tory 2019). Implementation of P^* in this way considers similar information to that considered at nodes 1 and 9 of the BUFO model (Figure 4.13) but omits explicit assessment of fuel availability, fire dynamics, and the likelihood of deep flaming.

4.4 Modeling Pyroconvective Interactions

Current operational fire spread models are mostly empirical, or quasi-empirical, and are designed to describe the behavior of a fire that has attained quasi-equilibrium. Pyroconvective interactions certainly play a significant role in the propagation of such fires (Finney et al. 2015) and so some pyroconvective effects are implicitly accounted for within current quasi-steady operational models through calibration of the model's parameters. However, pyroconvective interactions are not explicitly accounted for in a way that enables the current suite of operational models to faithfully represent dynamic fire line interactions or other forms of non-steady fire spread (Albini 1982; Dold and Zinoviev 2009; Dold 2011; Zekri et al. 2016). Indeed, until recently, full physical models or semi-physical models were the only type of models able to explicitly account for pyroconvective effects in wildfire propagation.

4.4.1 Full Physical Models and Semi-physical Coupled Fire–Atmosphere Models

As described by Sullivan (2009), a full physical model of fire behavior accounts for both the physical and chemical processes that drive fire spread (see also Chapter 1). This includes the thermal decomposition (especially pyrolysis) of the fuels, and the combustion process itself, occurring as a series of irreversible chemical reactions. These reactions are governed by the laws of chemistry, which can be used to describe changes in chemical species and, most importantly, the amount of heat that is released or absorbed. Subsequent transport of heat into unburnt fuel is governed by physical laws and described by equations that correspond to the different heat transfer mechanisms of radiation, conduction, and convection. A full physical model, therefore, involves grouping and coupling all the relevant mathematical equations and implementing a scheme – usually involving sophisticated numerical and computational techniques, to obtain a solution that reflects the behavior of a fire. As part of this process, convective heating is allowed to influence the motion of the air, which then feeds back on the behavior of the fire. Fully physical models such as WFDS and FIRETEC (Linn et al. 2002; Mell et al. 2013) can be used to model dynamic fire propagations down to sub-meter scales (e.g. Linn and Cunningham 2005).

While these models offer a level of physical realism in accounting for pyroconvective interactions, the computational cost of their implementation means that operational use is generally not feasible. For example, simulation of a small wildfire over the order of a few hours using a full physical model can take days or weeks, even with the use of modern high-performance computing resources. This is in stark contrast to the time frames associated with operational imperatives, which require faster than real time simulation of wildfire spread to inform decision-making.

Semi-physical fire propagation models are designed to strike a balance between the physical realism of the model and its computational overheads by simplifying the way the chemical or physical processes are accounted for. For example, this could involve a simplified description of combustion chemistry and focusing on modeling the physical processes alone. So, instead of explicitly modeling the combustion reaction intensity, a semi-physical model might simply assume that flames burn with a prescribed temperature, or with an intensity determined via an empirical model for a fire's rate of spread. This is the approach taken with models like CAWFE (Clark et al. 2004; Coen 2013; Coen and Schroeder 2017; Coen et al. 2020), WRF-Fire and WRF-SFire (Coen et al. 2013; Kochanski et al. 2013; Mandel et al. 2014) and ACCESS-Fire (Peace et al. 2018; Toivanen et al. 2019), which couple an empirically-based fire spread model with a numerical weather prediction model. These coupled fire–atmosphere models are generally used to simulate wildfire propagation at spatial scales of tens to hundreds of meters.

Both full physical and coupled fire–atmosphere models have been used successfully to account for pyroconvective effects in numerical experiments. Clark et al. (1996a, 1996b) used a coupled fire–atmosphere model to simulate the evolution of a straight-line fire into a parabolic fire line, and the dynamic and convective fingering observed in real fires (see Section 4.2). Thomas et al. (2017) used WRF-Fire to simulate the dynamic behavior of junction fires seen in the laboratory and field experiments of Viegas et al. (2012) and Raposo et al. (2018) (see Section 4.2.1). Sutherland et al. (2020) used WFDS to investigate the effect of ignition protocols on fire-spread rates in simulated grass fires. These simulations specifically considered the effect of ignition protocols resulting in the formation of convex initial fire lines compared to those that resulted in concave geometries. The simulation results indicated that ignition protocols, which result in initially concave fire lines, should be avoided in field experimentation as such configurations can produce a surge in the forward rate of spread of the fire before it settles down to the quasi-steady rate of spread that is primarily of interest. These results are consistent with those of other studies relating to junction fires, although Cruz et al. (2020) present a number of cautionary remarks about the validity of the findings of Sutherland et al. (2020). The vorticity-driven lateral spread observed in real fires (Sharples et al. 2012; Quill and Sharples 2015) has been modeled using WRF-Fire by Simpson et al. (2013, 2014, 2016); see Section 4.2.5, while eruptive fire behavior has been modeled using a static heat source in a computational fluid dynamics model by Edgar et al. (2015a, 2015b, 2016), extending the pioneering efforts of Simcox et al. (1992). This list of examples highlighting the utility of full physical and coupled fire–atmosphere models is by no means exhaustive.

4.4.2 Curvature-Based Models

In an effort to incorporate dynamic effects in a faster-than-real-time (FRT) modeling framework, Sharples et al. (2013) introduced a curvature-based fire propagation model. Motivated by the observed behavior of junction fires, they postulated that the enhanced rates of spread associated with portions of the fire line with concave geometry (such as near the point of intersection of a junction fire) could be accounted for by making the rate of spread a function of the fire line curvature.

Specifically, Sharples et al. (2013) considered the evolution of a junction fire as a plane curvature flow (Sethian 1985; Osher and Sethian 1988), with the merged fire line described by a time-dependent curve, $\gamma(t)$, whose evolution is governed by the normal flow equation

$$\hat{\boldsymbol{n}} \cdot \frac{\partial \gamma}{\partial t} = 1 - \epsilon k,$$

where k is the plane curvature of γ, $\hat{\boldsymbol{n}}$ is its (outward pointing) unit normal vector, and ϵ is a model parameter describing the strength of the effect of fire line curvature on the rate of spread. This model was able to produce results that were in good qualitative agreement with the dynamic behavior of the laboratory-scale junction fires of Viegas et al. (2012). The model was also able to qualitatively reproduce the evolution of the velocity of the junction point of the fire that was observed in experiments.

Hilton et al. (2016) used a level set method (Sethian 1999) to extend the curvature-based model of Sharples et al. (2013) to incorporate the effects of ambient wind. They used the resulting model to simulate the behavior of larger-scale experimental grassland fires that burnt over 33 m × 33 m plots. Importantly, the experimental fires were ignited in such a way that the initial fire line was concave, resembling the initial V shape of a junction fire (Cruz et al. 2015; Sutherland et al. 2020). The performance of the curvature-based model indicated that the inclusion of some aspect of fire line geometry can lead to a quantitative improvement in the prediction of the propagation of a fire perimeter, compared to a purely quasi-steady modeling approach. Indeed, the cases considered indicated that strongly concave curvature of a fire perimeter results in rapid acceleration of the fire line, although it was unclear whether the effect was due to a greater level of direct flame contact with unburnt fuels, enhanced radiative heating, increased convective heat flux, or a combination of these factors.

While curvature-based approaches were able to reproduce the dynamic behavior of junction fires and other fire lines with concave features, it is clear that fire line curvature will not be able to account for the dynamic fire propagation observed in connection with other geometric configurations. For example, two initially straight parallel lines of fire will have zero curvature, and so the curvature-based model will not be able to reproduce the convex geometry of the two fire lines that emerges in real cases. Similarly, the convex, parabolic head that develops in a wind-driven fire originating from a straight-line fire cannot be accounted for by a model that incorporates the effect of fire line curvature alone.

Thomas et al. (2017) and Thomas (2019) further examined the correlation between rate of spread and fire line curvature in their coupled fire–atmosphere simulations of junction fires. They found no clear relationship between the local curvature of the fire line and the pointwise instantaneous rate of spread of the fire. In a larger-scale analysis, involving notions of average rates of spread and curvature, Thomas (2019) used WRF-Fire to simulate arc fires, in which there is a well-defined notion of both the average curvature of the fire line and the average rate of spread. He found that the rate of spread depended dramatically on the angular length of the arc, but not on its curvature (see Figure 4.5). These results confirmed that in general fire line curvature is not a good predictor of dynamic fire spread. The improvements to prediction afforded by the inclusion of fire line curvature in the cases considered by Sharples et al. (2013) and Hilton et al. (2016) appear to be due to the fact that in some instances the curvature of a field is related to the two-dimensional Laplacian, which can be used to approximate the indraft into a fire's plume (see Section 4.4.3). However, while the (negative) curvature of the fire line acts as a useful proxy for a limited number of dynamic fire propagation scenarios, it does not faithfully represent the physics driving non-steady fire spread in general.

4.4.3 Pyrogenic Potential Model

Although fire line curvature was not found to be a good predictor of dynamic fire propagation in general, its success in predicting the dynamic evolution of fires in certain cases provides clues that can be used to develop successor models. Likewise, insights provided by coupled fire–atmosphere model output can be used to inform the development of reduced complexity models – which are also referred to as *models of intermediate complexity*. Such models can still be classified as quasi-physical, although the physical components may have been refined to the bare minimum required to reproduce the key dynamic effects driving non-steady fire propagation.

Hilton et al. (2018) proposed such a model based on the notion that the near surface pyroconvective air flow around a fire can be modeled as a two-dimensional potential flow (Batchelor and Batchelor 2000). The major benefit of the model is in the reduction of a complex three-dimensional flow field to a two-dimensional flow field, significantly reducing the computational complexity and, hence, cost. Since this air flow is induced by the presence of the fire, the potential flow was termed the *pyrogenic potential* flow to distinguish it from the conventional fluid flow potential (Hilton et al. 2018).

The modeling approach assumes that the pyrogenic flow near the ground (at flame height) is purely horizontal until it reaches the fire, at which point it immediately transforms into a purely vertical flow within the plume of the fire. As such, the model considers the plume simply as a sink for the horizontal flow of air and ignores any of the additional complexities of plume dynamics, other than the assumption that the strength of the sink can be related to the speed of the vertical updraft within the plume. This assumption is commonly used in plume dynamics modeling and has been shown to closely match experimental observations (Morton et al. 1956). The assumption implies that a more intense fire, which produces a stronger updraft, will generate a stronger horizontal pyrogenic indraft in the vicinity of the fire.

Drawing upon the standard tenets of the theory of potential flow, and discounting solenoidal (rotational) flow, the pyrogenic potential can be modeled using a two-dimensional Poisson equation of the form

$$\nabla^2 \psi = \rho, \tag{4.2}$$

where ψ is a scalar potential that defines the (two-dimensional) pyrogenic indraft $\boldsymbol{u}_p = \nabla \psi$, and ρ is a source term representing the strength of the sink. If the (three-dimensional) pyrogenic flow is assumed to be incompressible, then it follows that the sink term is described by (Hilton et al. 2018):

$$\rho(\boldsymbol{x}) = \begin{cases} -\dfrac{\partial w}{\partial z}, & \boldsymbol{x} \in \Omega \\ 0, & \boldsymbol{x} \notin \Omega \end{cases},$$

where $\boldsymbol{x}$ denotes (two-dimensional) horizontal position, Ω represents the (two-dimensional) flaming region, and w is the vertical component of the plume updraft velocity. To parameterize the model, we write $v = -\partial_z w$ and note that the parameter v will depend on factors such as the fire line intensity, fuel combustion rate, heat release rate, or flame height.

Fire line intensity is formulated as a one-dimensional entity via Byram's equation (Byram 1959):

$$I(\boldsymbol{x}') = HF(\boldsymbol{x}')R(\boldsymbol{x}'),$$

where $\boldsymbol{x}' \in \Gamma$ denotes location along a one-dimensional fire line (curve) Γ, H is the heat of combustion, $F(\boldsymbol{x}')$ is the fuel load, and $R(\boldsymbol{x}')$ is the local rate of spread. To apply this formulation to the pyrogenic potential model, it must be extended to two dimensions. The two-dimensional realization of Byram's fire line intensity over a two-dimensional flaming region Ω can be determined as (Towers 2007; Hilton et al. 2018):

$$I(\boldsymbol{x}) = HF(\boldsymbol{x})R(\boldsymbol{x})\delta(\phi(\boldsymbol{x}))\|\nabla\phi(\boldsymbol{x})\|,$$

where $\phi(\boldsymbol{x})$ is the scalar distance from Γ to point $\boldsymbol{x} \in \Omega$, and $\delta(\cdot)$ denotes the Dirac delta function. The final form of the forcing term v is given by:

$$v(\boldsymbol{x}) = \alpha R(\boldsymbol{x})\delta_s(\phi(\boldsymbol{x}))\|\nabla\phi(\boldsymbol{x})\|, \qquad (4.3)$$

where α is a parameter that describes the relationship between the forcing and fire intensity, as well as the fuel parameters H and F, and δ_s is a smoothed rendering of the Dirac delta function (Hilton et al. 2018).

Once the forcing term v has been determined, the pyrogenic potential flow $\boldsymbol{u}_p$ can be calculated via solution of the Poisson equation (Eq. (4.2)). The pyrogenic potential flow can then be combined with the effect of the ambient winds to drive the subsequent evolution of the fire (over a pre-defined time step) using the level set equation:

$$\frac{\partial\phi}{\partial t} = s\|\nabla\phi\| + \big(\boldsymbol{u}(\lambda) + \boldsymbol{u}_p\big) \cdot \nabla\phi,$$

where s is the basic rate of spread in the absence of wind, and

$$\boldsymbol{u}(\lambda) = \begin{cases} \lambda(\hat{\boldsymbol{w}} \cdot \hat{\boldsymbol{n}})\hat{\boldsymbol{w}}, & \hat{\boldsymbol{w}} \cdot \hat{\boldsymbol{n}} > 0 \\ 0, & \hat{\boldsymbol{w}} \cdot \hat{\boldsymbol{n}} \le 0 \end{cases}$$

accounts for the advective effect of the ambient wind on fire propagation. Here $\hat{\boldsymbol{w}}$ is the unit vector in the direction of the wind, $\hat{\boldsymbol{n}}$ is the outward-pointing unit normal to the fire perimeter, and λ is a model parameter describing the strength of the advective effect of the wind on fire spread. Note that the distance $\phi(\boldsymbol{x})$ required in Eq. (4.3) is calculated automatically in the implementation of the level set method.

It is important to note that the rate of spread R used to determine the source term in Eq. (4.3) can also be defined in terms of the level set function ϕ. This in turn is used to calculate $\boldsymbol{u}_p$, which is then used to determine ϕ (and hence R) at the next time step. As such, the pyrogenic potential model constitutes a basic coupled fire–atmosphere model – the fire dynamics feedback onto the fire itself via the pyrogenic potential, which emulates the pyroconvective effect of the fire's plume. The pyrogenic potential model is therefore able to account for some level of pyroconvective interaction between different parts of a fire line and appears to work well using only a single parameter α to represent the coupling. Figure 4.14 compares simulations conducted using the pyrogenic model with those conducted without any pyrogenic feedback between separate fires.

The pyrogenic model was originally designed to model the upwards forcing of air in fire plumes, but the upward forcing of air by topography, such as a steep hill, has a functionally identical effect. As such, an additional benefit of the pyrogenic potential model is that it can be extended to account for basic topographic

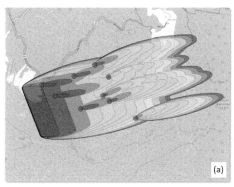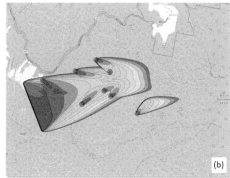

Figure 4.14 Fire spread simulations involving line and multiple point ignitions implemented in the Spark framework (Hilton et al. 2015; Miller et al. 2015): (a) Uncoupled simulation (i.e. without pyrogenic feedback) and (b) Simulation incorporating fire–atmosphere coupling using the pyrogenic potential model. In the uncoupled simulation, the fires spread independently of each other, while in the pyrogenic potential model simulation the fires interact with each other resulting in dynamic steering of the individual fires and a slowing of the overall progression of the fire complex.
A black and white version of this figure will appear in some formats. For the color version, please refer to the plate section.

correction of ambient winds at a negligible amount of additional computational cost. Hilton and Garg (2021) showed that this extension can create a terrain-corrected wind field almost identical to that of a three-dimensional diagnostic model in a fraction of the computational time.

Hilton et al. (2018) demonstrated the ability of the pyrogenic potential model to accurately reproduce the dynamic behavior of fires in laboratory and field-scale experiments. For example, the pyrogenic potential model was able to accurately account for the attraction between two individual wind-driven fires initiated from nearby point ignitions in a combustion tunnel, and the evolution of field-scale experimental fires initiated from V-shaped fire lines (Figure 4.15).

Thomas (2019) considered the pyrogenic inflow generated by a constant, static heat source in the shape of a circular disk with a radius of 100 meters. He compared the (radially symmetric) flow derived from analytical solution of Poisson's equation (as would apply in the pyrogenic potential model) to that simulated by the WRF atmospheric model. The results in Figure 4.16 show the radial profile of the pyrogenic inflow derived from the WRF model (at a height of 2 m) and from the analytic solution for two different source strengths. The first ($v = 0.076$) provides the best match to the peak strength of the pyrogenic inflow, although this peak occurs at a radius of 100 meters, as compared to the WRF generated flow which peaks at about half that radius. The second ($v = 0.02$) better matches the far-field behavior of the WRF model output but does a poor job of capturing the peak

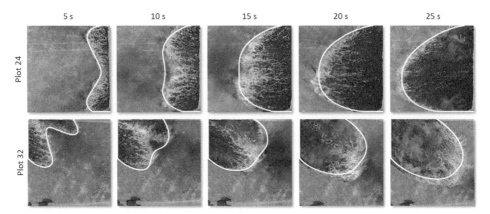

Figure 4.15 Comparison of simulations using the pyrogenic potential model to observations of the field-scale (30 m × 30 m) fire experiments of Cruz et al. (2015). The white lines show the simulated fire perimeter at each time step for plot 24 (top row) and plot 32 (bottom row).

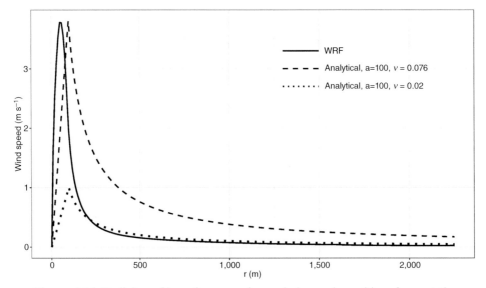

Figure 4.16 Radial profiles of near-surface wind speed resulting from static circular heat source of 100 meters radius (r is measured from the center of the heat source). The solid line is the WRF solution at 2 meters AGL, while the broken lines are analytical solutions of the Poisson equation (Eq. (4.2)) for different source strengths.

strength. The discrepancy between the two-dimensional analytical solutions and the three-dimensional coupled model output could be due to a variety of reasons; for example, nonlinear advection in the coupled model may act to squeeze the plume toward the center of the convective source. Overall, however, the WRF profile and the analytical solutions have the same general form, which suggests that further development of the pyrogenic model is warranted. One of the current challenges to further development of the pyrogenic model is understanding how the model parameters scale with fire size. At present, the parameter values used to accurately model small laboratory (10^{-1}–10^{0} meter) scale fires are an order of magnitude different to those used to model larger field experiment (10^{1}–10^{2} meter) scale fires (Hilton et al. 2018).

4.4.4 Near-Field Modeling of Pyroconvective Interactions

The pyrogenic potential model, in its original form, excluded consideration of any solenoidal flow that might be produced by the pyroconvective influence of the fire. However, the model can be extended to accommodate such effects through use of the Helmholtz decomposition, which states that a twice continuously differentiable vector field with compact support can be expressed as the sum of an irrotational (curl-free) vector field and a solenoidal (divergence-free) vector field (Arfken and Weber 1999). That is, if the pyrogenic flow $\boldsymbol{u}_p$ can be considered as relatively smooth and negligible at a considerable distance away from the fire (i.e. $\boldsymbol{u}_p \to 0$ as $r \to \infty$), then it can be decomposed as:

$$\boldsymbol{u}_p = \nabla \psi + \nabla \times \boldsymbol{\eta},$$

where ψ is a scalar potential (as in Eq. (4.2)) and $\boldsymbol{\eta}$ is a vector potential, which accounts for solenoidal flow. The scalar and vector potentials can be determined as solutions of the two-dimensional Poisson equations:

$$\nabla^2 \psi = \rho, \qquad \nabla^2 \boldsymbol{\eta} = \boldsymbol{\omega},$$

where ρ is defined as it was for Eq. (4.2), and $\boldsymbol{\omega}$ represents sources of vertical vorticity. Once the two potentials ψ and $\boldsymbol{\eta}$ have been determined, the effects of the fire on the local atmosphere can be accounted for by the pyrogenic flow $\boldsymbol{u}_p$. Since the Helmholtz decomposition only applies if the pyrogenic flow can be considered negligible sufficiently far away from the fire, this modeling framework only allows for incorporation of local influences of the fire on the atmosphere – we refer to these as *near-field effects* and refer to the model as the *near-field model* (Hilton et al. 2018; Sharples and Hilton 2020).

The fact that the near-field model can be used to incorporate the effects of vertical vorticity means that it can be used to model fire propagation in the presence of fire whirls (e.g. Figure 4.17) and to capture some of the dynamic effects exhibited in instances of VLS, for example. Indeed, Sharples and Hilton (2020) demonstrated how the near-field model can be used to reproduce the

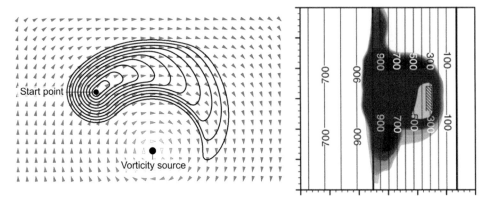

Figure 4.17 Near-field simulation of fire spread in the presence of a static vorticity source representing a fire whirl (left) and (right) a near-field simulation of VLS overlaid on output from a three-dimensional coupled fire–atmosphere model (ridge geometry with wind blowing from left to right across ridge; fire ignition on right, leeward, flank of ridge where horizontal vorticity is converted to vertical vorticity by the plume; grey scale indicating fuel fraction remaining).

patterns of fire propagation associated with VLS under several basic assumptions. Specifically, they combined the near-field model with a steady-state analytical model to account for the transformation of ambient horizontal vorticity into (pyrogenic) vertical vorticity. An example of how the two-dimensional near-field model output compares with that derived from a fully coupled, three-dimensional fire–atmosphere model can be seen on the right of Figure 4.17. It is of interest to note that, for a two-hour simulation, the near-field model ran for about 10–20 seconds, while the coupled fire–atmosphere model required about 10–12 hours of computation time. As such, the near-field model decreases the computational cost by about three orders of magnitude yet is still able to capture the main features of this dynamic mode of fire propagation; that is, the lateral spread along the ridge crest.

Hilton et al. (2019) further extended the near-field modeling of VLS by incorporating firebrands and spot fires. Firebrands were randomly generated, and their transport was simulated using a Lagrangian particle model. The extended near-field model was able to reproduce the main features of lateral spread and downwind spotting observed in real-world VLS events. While more research is required, the near-field modeling approach shows clear promise toward enabling more faithful real-time modeling of dynamic effects of phenomena like VLS and better operational capacity for prediction of deep flaming events. Figure 4.18 provides a schematic account of how a spot fire forming on a leeward slope can give rise to mass spotting and deep flaming via VLS.

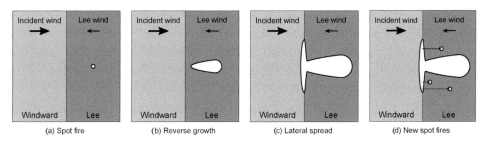

Figure 4.18 Schematic representation of how a spot fire on a leeward slope can produce deep flaming via VLS. (a) A spot fire ignites on a leeward slope, (b) the fire develops and starts to burn back up the slope, driven by winds that oppose the main incident wind direction, (c) the fire reaches the ridge crest and spreads laterally due to VLS, (d) the lateral spread zones act as a source of new embers, which start new spot fires. The process then repeats and compounds itself (see also Figure 4.9).

4.5 Discussion and Conclusions

Pyroconvection is a key factor driving the propagation of wildfires. Indeed, convection is the dominant heat transfer mechanism involved in the ignition of porous wildland fuels (Finney et al. 2015), which are unlikely to be ignited by radiant heat alone. Wildland fuels exposed to radiant heating will not generally rise to ignition temperature, due to the effect of convective cooling (Liu et al. 2015), whereas wildland fuels that are immersed in a hot, convective flow will ignite more readily. Finney et al. (2015) demonstrated that even at the most fundamental level, fuel elements are brought to ignition temperature through intermittent immersion in hot, convective flow (i.e. flames), interspersed with periods of convective cooling, with radiation playing only a relatively minor role in heating the fuels. In this sense, wildland fires can be viewed as quite distinct to structural fires, where radiant heating is typically the dominant mechanism propagating the fire and is a key factor in initiating extreme fire behaviors such as compartment flashover (e.g. Graham et al. 1995).

In addition to the fundamental role pyroconvection plays in basic wildfire propagation (even in the case of quasi-steady spread), this chapter has highlighted the critical importance of pyroconvective interactions in driving dynamic fire behaviors and the more general role they play in shaping the perimeter of a wildfire (e.g. the parabolic shape of a wind-driven head fire). Moreover, certain fire shapes will enhance the influence that pyroconvective interactions have on fire spread, such as is the case in the accelerated spread of junction fires, or the enhanced likelihood of fire whirl formation in the presence of certain fire line geometries. Pyroconvection can also drive dynamic interactions between the fire and the terrain, as is the case in eruptive fire behavior, or between the fire and terrain-modified winds, as is the case in vorticity-driven lateral spread.

In addition to the extreme fire behaviors just discussed, pyroconvective interactions and fire line geometry are also important considerations for prescribed burning practices. These include strategic hazard reduction burning during milder fire weather conditions, tactical backburning undertaken as part of wildfire suppression, and even design and implementation of wildfire experiments. For many decades, experienced fire practitioners have taken advantage of pyroconvective interactions arising from specific ignition patterns to assist in achieving desired fire behavioral prescriptions during hazard reduction burning, even in marginal conditions (e.g. Rothermel 1984). Alternatively, unexpected escalation in fire behavior driven by pyroconvective interactions can significantly compromise the objectives of hazard reduction burning (e.g. Densmore 2019).

During wildfire suppression operations, the pyroconvective influence of the main fire's plume can be used to draw backburns toward the main fire, ensuring that the backburn has the intended effect of burning out fuels ahead of the main fire, rather than becoming part of a larger wildfire complex. Sutherland et al. (2020) also highlight how certain ignition protocols can produce unintended dynamic surging in the fire's progression and can so compromise the aims of fire behavior experiments, especially those designed to inform quasi-steady model development.

Modes of fire propagation driven by pyroconvective interactions pose a number of challenges for fire behavior modeling. Traditional fire behavior models, which implicitly assume quasi-steady fire behavior, are, by their very design, ill-suited to describing non-steady behaviors. Indeed, functions that deliver a unique estimate of a fire's rate of spread based on ambient environmental conditions could be dangerously misleading in situations where dynamic feedbacks create local conditions that are significantly different to ambient conditions. This will be particularly problematic when dynamic effects result in rates of fire spread and intensities that are significantly higher than those predicted by traditional models. Viegas et al. (2009) present a number of critical firefighter safety incidents in which dynamic fire behaviors can be inferred, while Lahaye et al. (2017) explicitly discuss the safety implications of dynamic fire behaviors.

Several studies have shown how pyroconvective interactions can be at least partially accounted for using full physical models and semi-physical coupled fire–atmosphere models. While these modeling approaches provide valuable insights into the physical processes underpinning dynamic fire propagation, their computational expense means that they are generally not suitable for informing fire operations where faster-than-real-time simulation is preferred. This situation is likely to improve in the future with advances in computational capability; indeed, there are already a few recent examples of where such models have been used to provide enhanced operational guidance (Mandel et al. 2014; Coen and Schroeder 2017), but in the meantime there is a need for models of intermediate complexity that can capture important dynamic processes at minimal computational expense. Another issue with using full physical models and semi-physical models to account for dynamic fire propagation at landscape scales is the spatial resolution required to resolve critical features, such as the vertical vorticity responsible for vorticity-driven lateral spread (Simpson et al. 2014). Adequately resolving these types of features often requires grid resolutions of less than 100 meters, which can sometimes test the limits of these model's capabilities.

Dynamic fire behaviors driven by pyroconvective interactions have been implicated in the escalation of rate of spread and intensity associated with blow-up fire events and the development of extreme wildfires. Given the disproportionate impacts of these events, there is a clear mandate to continue model development and other research that improves our understanding of dynamic fire behaviors and our ability to predict its occurrence and the influence it has on fire development. This is particularly important given the apparent change in frequency of extreme fire weather events (Hasson et al. 2009; Boer et al. 2017) and the demonstrated link between broad-scale fuel dryness and severe fire activity (Nolan et al. 2016, 2020). Landscapes can transition to high levels of flammability due to the combined influence of significant, broad-scale rainfall and soil moisture deficits producing near maximum fuel availability, and critical fire weather events producing critically low fuel moisture content. Under these conditions, there is an increase in the likelihood that dynamic fire behaviors will develop into deep flaming events. This in turn increases the likelihood of extreme wildfire development, as was exemplified during the 2019/20 fire season in southeast Australia, which occurred at the end of Australia's warmest and driest year on record (Abram et al. 2021) and during which multiple fires had catastrophic impacts (Filkov et al. 2020b). Indeed, the number of extreme wildfires that developed into pyrocumulonimbus during the 2019/20 season was significantly greater than that experienced in any previous season.

The emerging realities of climate change and its implications for global wildfire present significant challenges for wildfire risk management. Based on an analysis of surface fire weather and lower tropospheric conditions, Di Virgilio et al. (2019) reported that the potential for extreme wildfires will increase under climate change. While there remain varying degrees of uncertainty about how other drivers of wildfire risk might be affected by climate change (Abram et al. 2021), demographic factors such as continued (sub)urban expansion into wildland areas will further confound risk management, especially as it relates to protection of lives, houses, and other assets in and around the wildland–urban interface. At present, it can be argued that mitigation measures such as building codes do not adequately consider the potential effects of dynamic fire behaviors and pyroconvective interactions. For example, there is currently no consideration of pyrogenic winds in the Australian Standard for building in bushfire prone areas, and the "design fire" concept, which is used to quantity the standard to which structures should be built, is predicated on quasi-steady fire behavior models. There is also a glaring need to continue to raise awareness amongst firefighters of the full potential of dynamic fire behavior and extreme wildfire development, and how it can effect operational imperatives, including those related to fire suppression, structural protection, and maintaining options for safe egress.

This chapter has presented a synopsis of some of the latest developments in our understanding of pyroconvective interactions, their links to fire geometry, and their role in driving dynamic fire behavior and extreme wildfire development. It has highlighted the need to augment traditional quasi-steady wildfire modeling paradigms with more sophisticated approaches that combine highly instrumented, larger-scale experimental studies with state-of-the-art computational modeling. There is also a need to take maximum advantage of technical advances in remote sensing technology to provide new ways of observing extreme fire events. These include multispectral airborne scanners (e.g. line scans and FLIR systems), doppler radar (weather radar and portable systems), and satellite platforms. Data from these sources can be used to develop more sophisticated models for wildfire propagation, including stochastic models that explicitly incorporate system uncertainties and quantify "rate of spread" in a distributional sense, rather than as a single numerical value. More comprehensive understanding of pyroconvective interactions and dynamic fire behaviors can also form the basis for more targeted prescribed burning practices, more timely and accurate community warnings, and even for enhanced community resilience to wildfire impacts through urban design principles. These aspects of wildfire risk mitigation will become increasingly important for dealing with large wildland and wildland–urban interface fires under likely future climate scenarios.

References

Abram, NJ, Henley, BJ, Gupta, AS, Lippmann, TJR, Clarke, H, Dowdy, AJ, Sharples, JJ, Nolan, RH, Zhang, T, Wooster, MJ, Wurtzel, JB, Meissner, KJ, Pitman, AJ, Ukkola, AM, Murphy, BP, Tapper, NJ, Boer, MM (2021) Connections of climate change and variability to large and extreme forest fires in southeast Australia. *Communications Earth & Environment* **2**(1), 1–17.
AFAC (2012) *Bushfire Glossary*, Melbourne, Australia: AFAC Limited.
Albini, FA (1981) A model for the wind-blown flame from a line fire. *Combustion and Flame* **43**, 155–174.

Albini, FA (1982) Response of free-burning fires to nonsteady wind. *Combustion Science and Technology* **29**(3–6), 225–241.

Alexander, ME (1985) Estimating the length-to-breadth ratio of elliptical forest fire patterns. In Proceedings of the Eighth Conference on Fire and Forest Meteorology, April 29–May 2, Detroit, Michigan. Society of American Foresters, Bethesda, MD, pp. 287–304.

Anderson, DH, Catchpole, EA, De Mestre, NJ Parkes, T (1982) Modelling the spread of grass fires. *The Journal of the Australian Mathematical Society. Series B. Applied Mathematics* **23**(4), 451–466.

Arfken, GB, Weber, HJ (1999) *Mathematical Methods for Physicists*. San Diego, CA: AAPT.

Attiwill, PM, Adams, MA (2013) Mega-fires, inquiries and politics in the eucalypt forests of Victoria, south-eastern Australia. *Forest Ecology and Management* **294**, 45–53.

Attiwill, PM, Binkley, D (2013) Exploring the mega-fire reality 2011: A forest ecology and management conference, Florida State University Conference Center, Florida, USA, 14-17 November 2011. *Forest Ecology and Management* **294**, 1–261.

Badlan, RL, Sharples, JJ, Evans, JP, McRae, R (2017) The role of deep flaming in violent pyroconvection. In 22nd International Congress on Modelling and Simulation, December 3–8, Hobart, Tasmania, Australia.

Badlan, RL, Sharples, JJ, Evans, JP, McRae, R (2019) Insights into the role of fire geometry and violent pyroconvection. In 23rd International Congress on Modelling and Simulation, December 1–6, Canberra, ACT, Australia.

Badlan, RL, Sharples, JJ, Evans, JP, McRae, R (2021a) Factors influencing the development of violent pyroconvection. Part I: Fire size and stability. *International Journal of Wildland Fire* **30**(7), 484–497.

Badlan, RL, Sharples, JJ, Evans, JP, McRae, R (2021b) Factors influencing the development of violent pyroconvection. Part II: Fire geometry and intensity. *International Journal of Wildland Fire* **30**(7), 498–512.

Batchelor, CK, Batchelor, GK (2000) *An Introduction to Fluid Dynamics*. Cambridge: Cambridge University Press.

Blanchi, R, Leonard, J, Leicester, RH (2006) Bushfire risk at the rural/urban interface. In *Proceedings of the Australasian Bushfire Conference*, June 6–9, Brisbane.

Boer, MM, Nolan, RH, De Dios, VR, Clarke, H, Price, OF, Bradstock, RA (2017) Changing weather extremes call for early warning of potential for catastrophic fire. *Earth's Future* **5**(12), 1196–1202.

Briggs, GA (1975) Plume rise predictions. In Haugen, DA, ed. *Lectures on Air Pollution and Environmental Impact Analyses*. Boston: American Meteorological Society, pp. 59–111.

Butler, BW (1998) Fire Behavior Associated with the 1994 South Canyon Fire on Storm King Mountain, Colorado (No. 9). Research Paper RMRS-RP-9, US Department of Agriculture, Forest Service, Rocky Mountain Research Station, Ogden, UT.

Byram, GM (1959) Combustion of forest fuels. In: Davis, KP, ed. *Forest Fire: Control and Use*. New York: McGraw-Hill, pp. 61–89.

Canfield, JM, Linn, RR, Sauer, JA, Finney, M, Forthofer, J (2014) A numerical investigation of the interplay between fireline length, geometry, and rate of spread. *Agricultural and Forest Meteorology* **189**, 48–59.

Chen, K, McAneney, J (2004) Quantifying bushfire penetration into urban areas in Australia. *Geophysical Research Letters* **31**(12), L12212.

Cheney, NP, Gould, JS, McCaw, WL (2001) The dead-man zone: A neglected area of firefighter safety. *Australian Forestry* **64**(1), 45–50.

Cheney, NP, Gould, JS, McCaw, WL, Anderson, WR (2012) Predicting fire behaviour in dry eucalypt forest in southern Australia. *Forest Ecology and Management* **280**, 120–131.

Clark, TL, Coen, JL, Latham, D (2004) Description of a coupled atmosphere–fire model. *International Journal of Wildland Fire* **13**(1), 49–63.

Clark, TL, Jenkins, MA, Coen, JL, Packham, DR (1996a) A coupled atmosphere–fire model: Role of the convective Froude number and dynamic fingering at the fireline. *International Journal of Wildland Fire* **6**(4), 177–190.

Clark, TL, Jenkins, MA, Coen, JL, Packham, DR (1996b) A coupled atmosphere fire model: Convective feedback on fire-line dynamics. *Journal of Applied Meteorology* **35**(6), 875–901.

Coen, JL (2013) Modeling Wildland Fires: A Description of the Coupled Atmosphere–Wildland Fire Environment Model (CAWFE). NCAR Technical Note NCAR/TN-500+ STR. Boulder, CO.

Coen, JL, Cameron, M, Michalakes, J, Patton, EG, Riggan, PJ, Yedinak, KM (2013) WRF-Fire: Coupled weather–wildland fire modeling with the weather research and forecasting model. *Journal of Applied Meteorology and Climatology* **52**(1), 16–38.

Coen, JL, Schroeder, W (2017) Coupled weather–fire modeling: From research to operational forecasting. *Fire Management Today* **75**(1), 39–45.

Coen, JL, Schroeder, W, Conway, S, Tarnay, L (2020) Computational modeling of extreme wildland fire events: A synthesis of scientific understanding with applications to forecasting, land management, and firefighter safety. *Journal of Computational Science*, **45**, 101152.

Countryman, CM (1971) Fire Whirls, Why, When, and Where. Pacific Southwest Forest and Range Experiment Station, US Forest Service, Berkeley, CA.

Cruz, MG, Alexander, ME (2017) Modelling the rate of fire spread and uncertainty associated with the onset and propagation of crown fires in conifer forest stands. *International Journal of Wildland Fire* **26**(5), 413–426.

Cruz, MG, Gould, JS, Kidnie, S, Bessell, R, Nichols, D, Slijepcevic, A (2015) Effects of curing on grassfires: II. Effect of grass senescence on the rate of fire spread. *International Journal of Wildland Fire* **24**(6), 838–848.

Cruz, MG, McCaw, WL, Anderson, WR, Gould, JS (2013) Fire behaviour modelling in semi-arid mallee-heath shrublands of southern Australia. *Environmental Modelling & Software* **40**, 21–34.

Cruz, MG, Sullivan, AL, Bessell, R, Gould, JS (2020) The effect of ignition protocol on the spread rate of grass fires: A comment on the conclusions of Sutherland et al. (2020). *International Journal of Wildland Fire* **29**(12), 1133–1138.

Cruz, MG, Sullivan, AL, Gould, JS, Sims, NC, Bannister, AJ, Hollis, JJ, Hurley, RJ (2012) Anatomy of a catastrophic wildfire: The Black Saturday Kilmore East fire in Victoria, Australia. *Forest Ecology and Management* **284**, 269–285.

Cunningham, P, Goodrick, SL, Hussaini, MY, Linn, R (2005) Coherent vortical structures in numerical simulations of buoyant plumes from wildland fires, *International Journal of Wildland Fire* **14**(1), 61–75.

Cunningham, P, Reeder, MJ (2009) Severe convective storms initiated by intense wildfires: Numerical simulations of pyro-convection and pyro-tornadogenesis. *Geophysical Research Letters* **36**(12), L12912.

Densmore, VS (2019) Key factors contributing to two pyrocumulonimbus clouds erupting during a prescribed burn in Western Australia. *6th International Fire Behaviour and Fuels Conference*, Sydney, April–May 2019.

Di Virgilio, G, Evans, JP, Blake, SA, Armstrong, M, Dowdy, AJ, Sharples, JJ, McRae, R (2019) Climate change increases the potential for extreme wildfires. *Geophysical Research Letters* **46**(14), 8517–8526.

Dold, J (2011) Fire spread near the attached and separated flow transition, including surge and stall behaviour. In *Proceedings of the 19th International Congress on Modelling and Simulation,* December, Perth.

Dold, J, Scott, K, Sanders, J (2011) The processes driving an ember storm. In *Proceedings of the 19th International Congress on Modelling and Simulation*, December, Perth.

Dold, JW, Zinoviev, A (2009) Fire eruption through intensity and spread rate interaction mediated by flow attachment. *Combustion Theory and Modelling* **13**(5), 763–793.

Drysdale, DD, Macmillan, AJR, Shilitto, D (1992) The King's Cross fire: Experimental verification of the "Trench effect." *Fire Safety Journal* **18**(1), 75–82.

Edgar, RA, Sharples, JJ, Sidhu, HS (2015a) Investigation of flame attachment and accelerated fire spread. In Asia Pacific Confederation of Chemical Engineering Congress 2015: APCChE 2015, incorporating CHEMECA 2015. Engineers Australia. September 27–October 1, Melbourne, p. 507.

Edgar, RA, Sharples, JJ, Sidhu, HS (2015b) Revisiting the King's Cross underground disaster with implications for modelling wildfire eruption. In *Proceedings of the 21st International Congress on Modelling and Simulation*. November 29–December 4, Broadbeach, Queensland, pp. 215–221.

Edgar, RA, Sharples, JJ, Sidhu, HS (2016) Examining the effects of convective intensity on plume attachment in three-dimensional trenches. In: *Chemeca 2016: Chemical Engineering–Regeneration, Recovery and Reinvention*. September 25–28, Adelaide. Melbourne: Engineers Australia, pp. 613–621.

Filkov, AI, Duff, TJ, Penman, TD (2020a) Frequency of dynamic fire behaviours in Australian forest environments. *Fire*, **3**(1), 1–17.

Filkov, AI, Ngo, T, Matthews, S, Telfer, S, Penman, TD (2020b) Impact of Australia's catastrophic 2019/20 bushfire season on communities and environment. Retrospective analysis and current trends. *Journal of Safety Science and Resilience* **1**(1), 44–56.

Finney, MA (1998) FARSITE: Fire Area Simulator: Model Development and Application. USDA Forest Service, Rocky Mountain Research Station Research Paper RMRS-RP-4. Ogden, UT.

Finney, MA, Cohen, JD, Forthofer, JM, McAllister, SS, Gollner, MJ, Gorham, DJ, Saito, K, Akafuah, NK, Adam, BA, English, JD (2015) Role of buoyant flame dynamics in wildfire spread. *Proceedings of the National Academy of Sciences* **112**(32), 9833–9838.

Finney, MA, McAllister, SS (2011) A review of fire interactions and mass fires. *Journal of Combustion* **2011**, 548328.

Forthofer, JM (2019) Fire tornadoes. *Scientific American* **321**(6), 60–67.

Forthofer, JM, Goodrick, SL (2011) Review of vortices in wildland fire. *Journal of Combustion* 2011, 984363.

Fromm, M, Bevilacqua, R, Servranckx, R, Rosen, J, Thayer, JP, Herman, J, Larko, D (2005) Pyro-cumulonimbus injection of smoke to the stratosphere: Observations and impact of a super blowup in northwestern Canada on 3–4 August 1998. *Journal of Geophysical Research: Atmospheres* **110**(D8), D08205.

Fromm, M, Lindsey, DT, Servranckx, R, Yue, G, Trickl, T, Sica, R, Doucet, P, Godin-Beekmann, S (2010) The untold story of pyrocumulonimbus. *Bulletin of the American Meteorological Society* **91**(9), 1193–1210.

Fromm, M, Shettle, EP, Fricke, KH, Ritter, C, Trickl, T, Giehl, H, Gerding, M, Barnes, JE, O'Neill, M, Massie, ST, Blum, U (2008a) Stratospheric impact of the Chisholm pyrocumulonimbus eruption: 2. Vertical profile perspective. *Journal of Geophysical Research: Atmospheres* **113**(D8), D08203.

Fromm, M, Torres, O, Diner, D, Lindsey, D, Vant Hull, B, Servranckx, R, Shettle, EP, Li, Z (2008b) Stratospheric impact of the Chisholm pyrocumulonimbus eruption: 1. Earth-viewing satellite perspective. *Journal of Geophysical Research: Atmospheres* **113**(D8), D08202.

Fromm, M, Tupper, A, Rosenfeld, D, Servranckx, R, McRae, R (2006) Violent pyro-convective storm devastates Australia's capital and pollutes the stratosphere. *Geophysical Research Letters* **33**(5), L05815.

Gill, AM (2005) Landscape fires as social disasters: an overview of "the bushfire problem." *Global Environmental Change Part B: Environmental Hazards* **6**(2), 65–80.

Glickman, TS (2000) *Glossary of Meteorology*, 2nd ed. Boston, MA: American Meteorological Society.

Goens, DW (1978) Fire Whirls. Missoula, MT: NOAA Technical Memorandum NWS WR-129.

Graham, TL, Makhviladze, GM, Roberts, JP (1995) On the theory of flashover development. *Fire Safety Journal*, **25**(3), 229–259.

Green, DG (1983) Shapes of simulated fires in discrete fuels. *Ecological Modelling* **20**(1), 21–32.

Green, DG, Gill, AM, Noble, IR (1983) Fire shapes and the adequacy of fire-spread models. *Ecological Modelling* **20**(1), 33–45.

Grumstrup, TP, McAllister, SS, Finney, MA (2017) Qualitative flow visualization of flame attachment on slopes. In *Proccedings of the 10th US National Combustion Meeting Organized by the Eastern States Section of the Combustion Institute*; April 23–26, 2017; College Park, MD. Pittsburgh, PA: The Combustion Institute.

Hasson, AEA, Mills, GA, Timbal, B, Walsh, K (2009) Assessing the impact of climate change on extreme fire weather events over southeastern Australia. *Climate Research* **39**(2), 159–172.

Hilton, JE, Garg, N, (2021) Rapid wind-terrain correction for wildfire simulations. *International Journal of Wildland Fire* **30**(6), 410–427.

Hilton, JE, Garg, N, Sharples, JJ (2019) Incorporating firebrands and spot fires into vorticity-driven wildfire behaviour models. In *23rd International Congress on Modelling and Simulation*. December 1–6, Canberra, ACT, Australia.

Hilton, JE, Miller, C, Sharples, JJ, Sullivan, AL (2016) Curvature effects in the dynamic propagation of wildfires. *International Journal of Wildland Fire*, **25**(12), 1238–1251.

Hilton, JE, Miller, C, Sullivan, AL, Rucinski, C (2015) Effects of spatial and temporal variation in environmental conditions on simulation of wildfire spread. *Environmental Modelling and Software* **67**, 118–127.

Hilton, JE, Sharples, JJ, Sullivan, AL, Swedosh, W (2017) Simulation of spot fire coalescence with dynamic feedback. In *22nd International Congress on Modelling and Simulation, Hobart, Tasmania*.

Hilton, JE, Sullivan, A, Swedosh, W, Sharples, J, Thomas, C (2018) Incorporating convective feedback in wildfire simulations using pyrogenic potential. *Environmental Modelling and Software* **107**, 12–24.

Kinniburgh, DC (2020) *Dynamics of Coupled Fire–Atmosphere Interactions*. PhD Thesis. Monash University. https://doi.org/10.26180/5f04329f6f2ef

Kochanski, AK, Jenkins, MA, Mandel, J, Beezley, JD, Clements, CB, Krueger, S (2013) Evaluation of WRF-SFIRE performance with field observations from the FireFlux experiment. *Geoscientific Model Development* **6**(4), 1109–1126.

Koo, E, Linn, RR, Pagni, PJ, Edminster, CB (2012) Modelling firebrand transport in wildfires using HIGRAD/FIRETEC. *International Journal of Wildland Fire* **21**(4), 396–417.

Koo, E, Pagni, PJ, Weise, DR, Woycheese, JP (2010) Firebrands and spotting ignition in large-scale fires. *International Journal of Wildland Fire* **19**(7), 818–843.

Kuwana, K, Sekimoto, K, Saito, K, Williams, FA, Hayashi, Y, Masuda, H (2007) Can we predict the occurrence of extreme fire whirls? *AIAA Journal* **45**(1), 16–19.

Lahaye, S, Curt, T, Fréjaville, T, Sharples, JJ, Paradis, L, Hély, C (2018a) What are the drivers of dangerous fires in Mediterranean France? *International Journal of Wildland Fire* **27**(3), 155–163.

Lahaye, S, Sharples, J, Matthews, S, Heemstra, S, Price, O (2017) What are the safety implications of dynamic fire behaviours? In: *22nd International Congress on Modelling and Simulation*. Decembere 3–8, Hobart, Tasmania , pp. 1125–1130.

Lahaye, S, Sharples, JJ, Matthews, S, Heemstra, S, Price, O, Badlan, R (2018b) How do weather and terrain contribute to firefighter entrapments in Australia? *International Journal of Wildland Fire* **27**(2), 85–98.

Lareau, NP, Clements, CB (2016) Environmental controls on pyrocumulus and pyrocumulonimbus initiation and development. *Atmospheric Chemistry and Physics* **16**(6), 4005–4022.

Lareau, NP, Clements, CB (2017) The mean and turbulent properties of a wildfire convective plume. *Journal of Applied Meteorology and Climatology* **56**(8), 2289–2299.

Lareau, NP, Nauslar, NJ, Abatzoglou, JT (2018) The Carr Fire vortex: A case of pyrotornadogenesis? *Geophysical Research Letters* **45**(23), 13–107.

Linn, R, Cunningham, P (2005) Numerical simulations of grass fires using a coupled atmosphere–fire model: Basic fire behavior and dependence on wind speed. *Journal of Geophysical Research* **110**(D13), D13107.

Linn, R, Reisner, J, Colman, JJ, Winterkamp, J (2002) Studying wildfire behavior using FIRETEC. *International Journal Wildland Fire* **11**(4), 233–246.

Liu N, Liu Q, Deng Z, Kohyu S, Zhu J (2007) Burn-out time data analysis on interaction effects among multiple fires in fire arrays. *Proceedings of the Combustion Institute*, **31**(2), 2589–2597.

Liu, N, Wu, J, Chen, H, Zhang, L, Deng, Z, Satoh, K, Viegas, DX, Raposo, J (2015) Upslope spread of a linear flame front over a pine needle fuel bed: The role of convection cooling. *Proceedings of the Combustion Institute*, **35**(3), 2691–2698.

Luderer, G, Trentmann, J, Winterrath, T, Textor, C, Herzog, M, Graf, HF, Andreae, MO (2006) Modeling of biomass smoke injection into the lower stratosphere by a large forest fire (Part II): sensitivity studies. *Atmospheric Chemistry and Physics* **6**(12), 5261–5277.

Luke RH, McArthur AG (1978) *Bushfires in Australia*. Canberra: AGPS.

Mandel, J, Amram, S, Beezley, JD, Kelman, G, Kochanski, AK, Kondratenko, VY, Lynn, BH, Regev, B, Vejmelka, M (2014) Recent advances and applications of WRF-SFIRE. *Natural Hazards and Earth System Sciences* **14**(10), 2829.

Martin, RE, Finney, MA, Molina, DM, Sapsis, DB, Stephens, SL, Scott, JH, Weise, DR (1991) Dimensional analysis of flame angles versus wind speed. In Andrews, PL, Potts, DF, eds. *Proceedings of the 11th Conference on Fire and Forest Meteorology*, April 16–19, 1991, Missoula, MT. Bethesda, MD: Society of American Foresters, pp. 212–217.

McArthur, AG (1967) *Fire Behaviour in Eucalypt Forests*. Leaflet 107. Canberra: Forestry and Timber Bureau.

McCarthy, N. (2020) Bushfire Thunderstorms: Radar Analysis of Fire-driven Convection in Australia. PhD Thesis. University of Queensland. https://doi.org/10.14264/uql .2020.704

McCarthy, N, McGowan, H, Guyot, A, Dowdy, A (2018) Mobile X-Pol radar: A new tool for investigating pyroconvection and associated wildfire meteorology. *Bulletin of the American Meteorological Society*, **99**(6), 1177–1195.

McRae, RHD, Sharples, JJ (2013) A process model for forecasting conditions conducive to blow-up fire events. In *Proceedings of the 2013 MODSIM Conference*, December 1–6, Adelaide, Australia.

McRae, RHD, Sharples, JJ (2014) Forecasting conditions conducive to blow-up fire events. *CAWCR Research Letters* (11), 14–19.

McRae, RHD, Sharples, JJ, Fromm, M (2015) Linking local wildfire dynamics to pyroCb development. *Natural Hazards and Earth Systems Science* **15**(3), 417–428.

McRae, RHD, Sharples, JJ, Wilkes, SR, Walker, A (2013) An Australian pyro-tornadogenesis event. *Natural Hazards* **65**, 1801–1811.

Mell, W, Charney, J, Jenkins, MA, Cheney, P, Gould, J (2013) Numerical simulations of grassland fire behavior from the LANL-FIRETEC and NIST-WFDS models. In: Qu, JJ, Sommers, WT, Yang, R, Riebau, AR, eds. *Remote Sensing and Modeling Applications to Wildland Fires*. Berlin: Springer, pp. 209–225.

Miller, C, Hilton, J, Sullivan, A, Prakash, M (2015) SPARK: A bushfire spread prediction tool. In Denzer, R, Argent, RM, Schimak, G, Hřebíček, J, eds. *Environmental Software Systems. Infrastructures, Services and Applications: 11th IFIP WG 5.11 International Symposium, ISESS 2015, Melbourne, VIC, Australia, March 25–27, 2015, Proceedings*, Vol. 448. Cham: Springer, pp. 262–271.

Mills, GA, McCaw, WL (2010) *Atmospheric stability environments and fire weather in Australia: extending the Haines Index*. CAWCR Technical report No. 20, March 2010.

Morton, BR, Taylor, GI, Turner, JS (1956) Turbulent gravitational convection from maintained and instantaneous sources. *Proceedings of the Royal Society of London. Series A. Mathematical and Physical Sciences* **234**(1196), 1–23.

Ndalila, MN, Williamson, GJ, Fox-Hughes, P, Sharples, J, Bowman, DM (2020) Evolution of a pyrocumulonimbus event associated with an extreme wildfire in Tasmania, Australia. *Natural Hazards and Earth System Sciences* **20**(5), 1497–1511.

Nolan, RH, Blackman, CJ, de Dios, VR, Choat, B, Medlyn, BE, Li, X, Bradstock, RA, Boer, MM (2020) Linking forest flammability and plant vulnerability to drought. *Forests* **11**(7), 779.

Nolan, RH, Boer, MM, Resco de Dios, V, Caccamo, G, Bradstock, RA (2016) Large-scale, dynamic transformations in fuel moisture drive wildfire activity across southeastern Australia. *Geophysical Research Letters* **43**(9), 4229–4238.

NWCG (2012) *Glossary of Wildland Fire Terminology*. National Wildfire Coordinating Group, Incident Operations Standards Working Team. www.nwcg.gov/glossary/a-z, Last accessed November 23, 2021.

Osher, S, Sethian, JA (1988) Fronts propagating with curvature-dependent speed: Algorithms based on Hamilton-Jacobi formulations. *Journal of Computational Physics* **79**(1), 12–49.

Page, WG, Butler, BW (2017) An empirically based approach to defining wildland firefighter safety and survival zone separation distances. *International Journal of Wildland Fire* **26**(8), 655–667.

Peace, M, Kepert, JD, Ye, H (2018) The ACCESS-Fire coupled fire-atmosphere model. In 33rd Conference on Agricultural and Forest Meteorology/12th Fire and Forest

Meteorology Symposium/Fourth Conference on Biogeosciences. American Meteorological Society.

Peterson, DA, Hyer, EJ, Campbell, JR, Solbrig, JE, Fromm, MD (2017) A conceptual model for development of intense pyrocumulonimbus in western North America. *Monthly Weather Review* **145**(6), 2235–2255.

Pinto, C, Viegas, D, Almeida, M, Raposo, J (2017) Fire whirls in forest fires: An experimental analysis. *Fire Safety Journal* **87**, 37–48.

Porterie, B, Zekri, N, Clerc, JP, Loraud, JC (2005) Influence des brandons sur la propagation d'un feu de forêt. *Comptes Rendus Physique*, **6**(10), 1153–1160.

Potter, BE (2005) The role of released moisture in the atmospheric dynamics associated with wildland fires. *International Journal of Wildland Fire* **14**(1), 77–84.

Potter, BE (2012a) Atmospheric interactions with wildland fire behaviour: I. Basic surface interactions, vertical profiles and synoptic structures. *International Journal of Wildland Fire* **21**(7), 779–801.

Potter, BE (2012b) Atmospheric interactions with wildland fire behaviour: II. Plume and vortex dynamics. *International Journal of Wildland Fire* **21**(7), 802–817.

Quill, R, Sharples, JJ (2015) Dynamic development of the 2013 Aberfeldy fire. In *Proceedings of the 21st International Congress on Modelling and Simulation.* November 29–December 4, Broadbeach, Queensland, pp. 284–290.

Raposo, J, Cabiddu, S, Viegas, D, Salis, M, Sharples, J (2015) Experimental analysis of fire spread across a two-dimensional ridge under wind conditions. *International Journal of Wildland Fire* **24**(7), 1008–1022.

Raposo, J, Viegas, D, Xie, X, Almeida, M, Figueiredo, A, Porto, L, Sharples, J (2018) Analysis of the physical processes associated with junction fires at laboratory and field scales. *International Journal Wildland Fire* **27**(1), 52–68.

Raposo, J (2016) Extreme Fire Behaviour Associated with the Merging of Two Linear Fire Fronts. Doctoral dissertation. Coimbra. Tese de doutoramento. http://hdl.handle.net/10316/31020

Rothermel, RC (1972) A Mathematical Model for Predicting Fire Spread in Wildland Fuels, Research Paper INT-115. U.S. Department of Agriculture, Intermountain Forest and Range, Ogden, UT.

Rothermel, RC (1984) Fire behavior consideration of aerial ignition. In *Proceedings of the Prescribed Fire by Aerial Ignition, Proceedings of a Workshop*, October 30–November 1, Missoula, MT, USA (Vol. 30).

SCFAIT (1994) *Report of the South Canyon Fire Investigation Team: South Canyon Fire Investigation of the 14 Fatalities that Occurred on July 6, 1994 near Glenwood Springs, Colorado.* Washington, DC: US Forest Service and Bureau of Land Management.

Sethian, JA (1985) Curvature and the evolution of fronts. *Communications in Mathematical Physics*, **101**, 487–499.

Sethian, JA (1999) *Level Set Methods and Fast Marching Methods: Evolving Interfaces in Computational Geometry, Fluid Mechanics, Computer Vision, and Materials Science.* Cambridge: Cambridge University Press.

Sharples, JJ, Cary, GJ, Fox-Hughes, P, Mooney, S, Evans, JP, Fletcher, MS, Fromm, M, Grierson, PF, McRae, R, Baker, P (2016) Natural hazards in Australia: Extreme bushfire. *Climatic Change* **139**, 85–99.

Sharples, JJ, Cechet, RP (2017) Reassessing the validity of AS3959 in the presence of dynamic bushfire propagation. In: *22nd International Congress on Modelling and Simulation.* December 3–8, Hobart, Tasmania.

Sharples, JJ, Gill, AM, Dold, JW (2010) The trench effect and eruptive wildfires: Lessons from the King's Cross Underground disaster. In *Proceedings of Australian Fire and*

Emergency Service Authorities Council 2010 Conference. September 8–10, Darwin, Australia, Vol. 2010, pp. 8–10.

Sharples, JJ, Hilton, JE (2020) Modeling vorticity-driven wildfire behavior using near-field techniques. *Frontiers in Mechanical Engineering*, **5**, 69.

Sharples, JJ, Kiss, A, Raposo, J, Viegas, D, Simpson, C (2015) Pyrogenic vorticity from windward and lee slope fires. In *Proceedings of the 21st International Congress on Modelling and Simulation*. November 29–December 4, Broadbeach, Queensland, pp. 291–297.

Sharples, JJ, McRae, RHD (2011) Evaluation of a very simple model for predicting the moisture content of eucalypt litter. *International Journal of Wildland Fire* **20**(8), 1000–1005.

Sharples, JJ, McRae, RHD, Weber, RO, Gill, AM (2009) A simple index for assessing fuel moisture content. *Environmental Modelling & Software* **24**(5), 637–646.

Sharples, JJ, McRae, RHD, Wilkes, S (2012) Wind–terrain effects on the propagation of wildfires in rugged terrain: Fire channelling. *International Journal of Wildland Fire* **21**(3), 282–296.

Sharples, JJ, Towers, IN, Wheeler, G, Wheeler, VM, McCoy, JA (2013) Modelling fire line merging using plane curvature flow. In: *Proceedings of the 19th International Congress on Modelling and Simulation*, December 1–6, Adelaide.

Simcox, S, Wilkes, NS, Jones, IP (1992) Computer simulation of the flows of hot gases from the fire at King's Cross underground station. *Fire Safety Journal* **18**(1), 49–73.

Simpson, CC, Sharples, JJ, Evans, JP (2014) Resolving vorticity-driven lateral fire spread using the WRF-fire coupled atmosphere–fire numerical model. *Natural Hazards and Earth Systems Science* **14**(9), 2359–2371.

Simpson, CC, Sharples, JJ, Evans, JP (2016) Sensitivity of atypical lateral fire spread to wind and slope. *Geophysical Research Letters* **43**(4), 1744–1751.

Simpson, CC, Sharples, JJ, Evans, JP, McCabe, M (2013) Large eddy simulation of atypical wildland fire spread on leeward slopes. *International Journal of Wildland Fire* **22**(5), 599–614.

Storey, MA, Price, OF, Bradstock, RA, Sharples, JJ (2020a) Analysis of variation in distance, number, and distribution of spotting in southeast Australian wildfires. *Fire* **3**(2), 10.

Storey, MA, Price, OF, Sharples, JJ, Bradstock, RA (2020b) Drivers of long-distance spotting during wildfires in south-eastern Australia. *International Journal of Wildland Fire* **29**(6), 459–472.

Sullivan, AL (2009) Wildland surface fire spread modelling, 1990–2007. 1: Physical and quasi-physical models. *International Journal of Wildland Fire* **18**(4), 349–368.

Sullivan, AL, Swedosh, W, Hurley, RJ, Sharples, JJ, Hilton, JE (2019) Investigation of the effects of interactions of intersecting oblique fire lines with and without wind in a combustion wind tunnel. *International Journal of Wildland Fire* **28**(9), 704–719.

Sutherland, D, Sharples, JJ, Moinuddin, KA (2020) The effect of ignition protocol on grassfire development. *International Journal of Wildland Fire* **29**(1), 70–80.

Tedim, F, Leone, V, Amraoui, M, Bouillon, C, Coughlan, MR, Delogu, GM, Fernandes, PM, Ferreira, C, McCaffrey, S, McGee, TK, Parente, J (2018) Defining extreme wildfire events: difficulties, challenges, and impacts. *Fire* **1**(1), 9.

Thomas, CM (2019) *Investigation of Spotting and Intrinsic Fire Dynamics Using a Coupled Atmosphere–Fire Modelling Framework*. PhD Thesis. University of New South Wales.

Thomas, CM, Sharples, J. Evans, JP (2017) Modelling the dynamic behaviour of junction fires with a coupled atmosphere–fire model. *International Journal of Wildland Fire* **26**(4), 331–344.

Tohidi, A, Gollner, MJ, Xiao, H (2018) Fire whirls. *Annual Review of Fluid Mechanics* **50**, 187–213.

Toivanen, J, Engel, CB, Reeder, MJ, Lane, TP, Davies, L, Webster, S, Wales, S (2019) Coupled atmosphere–fire simulations of the Black Saturday Kilmore East wildfires with the Unified Model. *Journal of Advances in Modeling Earth Systems* **11**(1), 210–230.

Tolhurst, K, Shields, B, Chong, D (2008) Phoenix: Development and application of a bushfire risk management tool. *Australian Journal of Emergency Management* **23**(4), 47.

Tory, KJ (2019) Pyrocumulonimbus firepower threshold: A pyrocumulonimbus prediction tool. In: Bates, J, ed. *AFAC19 powered by INTERSCHUTZ Extended abstracts from the Bushfire and Natural Hazards CRC Research Forum. Australian Journal of Emergency Management*, Monograph No. 5, pp. 21–27.

Tory, KJ, Kepert, JD (2021) Pyrocumulonimbus firepower threshold: Assessing the atmospheric potential for pyroCb. *Weather and Forecasting* **36**(2), 439–456.

Towers, JD (2007) Two methods for discretizing a delta function supported on a level set. *Journal of Computational Physics* **220**(2), 915–931.

Trentmann, J, Luderer, G, Winterrath, T, Fromm, MD, Servranckx, R, Textor, C, Herzog, M, Graf, HF, Andreae, MO (2006) Modeling of biomass smoke injection into the lower stratosphere by a large forest fire (Part I): Reference simulation. *Atmospheric Chemistry and Physics* **6**(12), 5247–5260.

Turco, M, Jerez, S, Augusto, S, Tarín-Carrasco, P, Ratola, N, Jiménez-Guerrero, P, Trigo, RM (2019) Climate drivers of the 2017 devastating fires in Portugal. *Scientific Reports* **9**, 1–8.

Tymstra, C, Bryce, RW, Wotton, BM, Taylor, SW, Armitage, OB (2010) Development and structure of Prometheus: The Canadian wildland fire growth simulation model. Natural Resources Canada, Canadian Forest Service, Northern Forestry Centre, Information Report NOR-X-417. Edmonton, AB.

Vallis, GK (2017) *Atmospheric and Oceanic Fluid Dynamics: Fundamentals and Large-Scale Circulation*, 2nd ed. Cambridge: Cambridge University Press.

Van Wagner, CE (1977a) Conditions for the start and spread of crown fire. *Canadian Journal of Forest Research* **7**(1), 23–34.

Van Wagner, CE (1977b) Effect of slope on fire spread rate. *Canadian Forestry Service Bimonthly Research Notes* **33**(1), 7–8.

Viegas, DX (2006) Parametric study of an eruptive fire behaviour model. *International Journal of Wildland Fire* **15**(2), 169–177.

Viegas, DX, Pita, LP (2004) Fire spread in canyons. *International Journal of Wildland Fire* **13**(3), 253–274.

Viegas, DX, Raposo, J, Davim, DA, Rossa, CG (2012) Study of the jump fire produced by the interaction of two oblique fire fronts. part 1. Analytical model and validation with no-slope laboratory experiments. *International Journal of Wildland Fire* **21**(7), 843–856.

Viegas, DX, Simeoni, A, Xanthopoulos, G, Rossa, C, Ribeiro, LM, Pita, LP, Stipanicev, D, Zinoviev, A, Weber, R, Dold, J, Caballero, D (2009) *Recent Forest Fire Related Accidents in Europe*. Luxembourg: Office for Official Publications of the European Communities.

Weise, DR, Biging, GS (1996) Effects of wind velocity and slope on flame properties. *Canadian Journal of Forest Research* **26**(10), 1849–1858.

Werth, PA, Potter, BE, Alexander, ME, Cruz, MG, Clements, CB, Finney, MA, Forthofer, JM, Goodrick, SL, Hoffman, C, Jolly, WM, McAllister, SS (2011) Synthesis of Knowledge of Extreme Fire Behavior, Vol. 1. Gen. Tech. Rep. PNW-GTR-854. U.S. Department of Agriculture, Forest Service, Pacific Northwest Research Station. Portland, OR.

Williams, J. (2013) Exploring the onset of high-impact mega-fires through a forest land management prism. *Forest Ecology and Management* **294**, 4–10.

Wu, Y, Xing, HJ, Atkinson, G (2000) Interaction of fire plume with inclined surface. *Fire Safety Journal* **35**(4), 391–403.

Zekri, N, Harrouz, O, Kaiss, A, Clerc, JP, Viegas, XD (2016) Generalized Byram's formula for arbitrary fire front geometries. *International Journal of Thermal Sciences* **110**, 222–228.

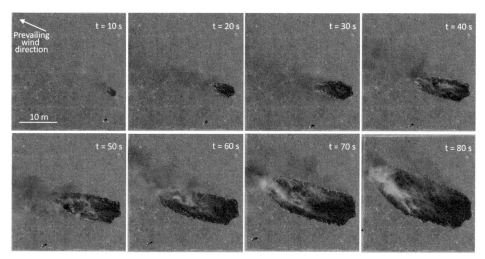

Figure 1.6 Time-series of aerial photographs of the growth of experimental fire Braidwood E14 in grass fuels ignited at a point in a 33×33 m plot, commencing 10 seconds after ignition. The growth of the fire in area, perimeter length, and forward rate of spread increases with time as the fire responds to slight changes in the direction of the wind. Conditions for experiment were: grass height 20 cm, grass curing 90%, dead fuel moisture content 7.5% (oven-dry weight), air temperature 301 K (28°C), relative humidity 30%, prevailing mean wind speed at 10 m 17.9 km h^{-1}, maximum interval rate of spread 2.16 km h^{-1}.

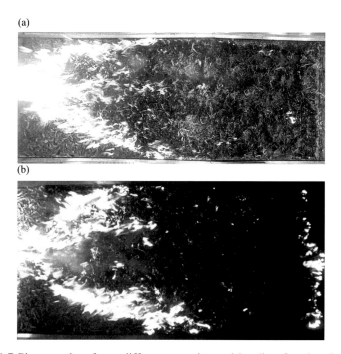

Figure 1.7 Photographs of two different experimental heading fires burning in dry eucalypt forest litter in the CSIRO Pyrotron combustion wind tunnel with wind blowing right to left. (a) Image taken 155 seconds after ignition. Most of the fuel bed has been consumed with only a small amount of residual flaming and two sources of smoldering combustion visible behind the flame front. (b) Image of another fire burning under similar conditions after the wind tunnel overhead lights have been turned off, revealing a multitude of glowing combustion behind the front.

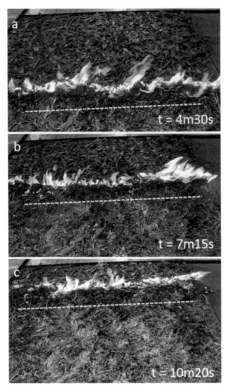

Figure 1.10 Time-series of oblique view photographs of the growth of an experimental flank fire burning in dry eucalypt litter in a 1.5 m s^{-1} wind and moisture content of 7–8% oven dry weight. The fire was ignited from a 1.5 meter-line on the parallel edge of a 1.5-meter fuel bed and had a mean spread rate of 6 m h^{-1} from bottom to top. Air flow is from left to right: (a) 4 minutes 30 seconds after ignition, (b) 7 minutes 15 seconds after ignition, (c) 10 minutes 20 seconds after ignition. The white dashed line roughly indicates the divide between the char zone undergoing oxidation and the ash zone that has completed oxidation.

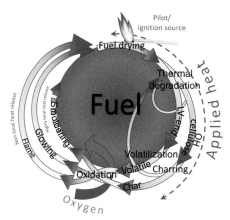

Figure 1.14 A revised version of the conceptual model presented in Figure 1.1 with the distance from the center of the circle indicating the magnitude of enthalpy. In this conceptual model, the fundamental competitive thermokinetics between the formation of levoglucosan-end cellulose (LV-end) and hydrolyzed cellulose (OH cellulose) during thermal degradation drives volatilization and the formation of volatile and charring and the formation of char, respectively. Heat formed in the production of hydrolyzed cellulose or char may be sufficient to drive the reaction toward the endothermic volatilization pathway. These thermal degradation products may then undergo oxidation in the form of flame in the case of volatile (which as a gas–gas reaction is open to turbulent mixing) and glowing or smoldering combustion in the case of char. The heat released by glowing or smoldering is highly localized within the fuel bed, while that released by flame may be some distance from the fuel bed.

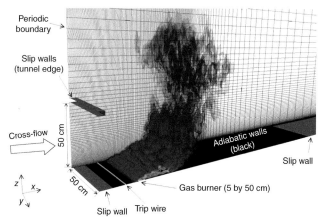

Figure 2.5 General view of the computational domain in LES simulations of the line fire configuration with wind. The burner is 5-cm deep in the x-direction and 50-cm wide in the y-direction; the wind tunnel at the inlet boundary of the computational domain is 50-cm wide in the y-direction and 50-cm high in the vertical z-direction. The flame is visualized using several isocontours of the instantaneous volumetric heat release rate.

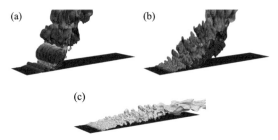

Figure 2.6 LES simulations of the change in flame/plume structure with wind velocity. Three-dimensional instantaneous structure of the plume for different values of the crosswind velocity: (a) $u_w = 0.75$ m s^{-1}; (b) $u_w = 1.5$ m s^{-1}; (c) $u_w = 3$ m s^{-1}. The plume is colored according to the local value of the x-velocity; red (white) corresponds to a low (high) value.

Figure 2.12 General view of the computational domain in LES simulations of the line fire configuration with slope. The computational mesh is tilted so that grid lines remain normal or parallel to the inclined surface. The flame is visualized using volume rendering of temperature.

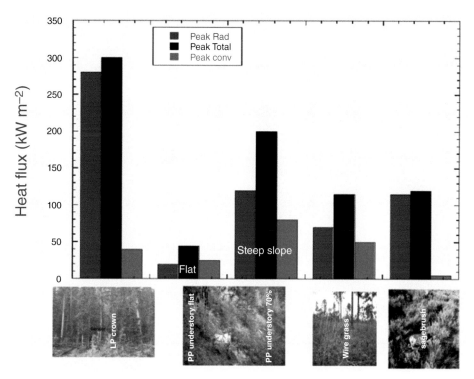

Figure 3.4 Time-averaged peak total, radiant, and convective energy from fires in various fuel types (Frankman et al. 2012).

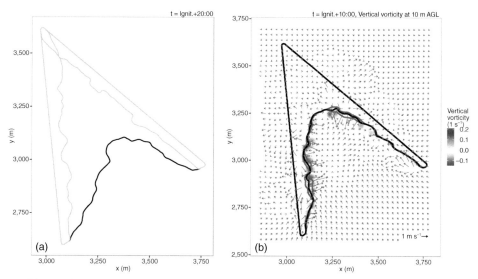

Figure 4.4 Junction fire behavior with $2\theta = 45°$, simulated using WRF-Fire (Thomas et al. 2017). (a) Grey lines represent the progression of two fire lines 20 minutes after ignition. The evolution of each of the two fire lines has been simulated separately and in the absence of the other. The black line shows the position of the simulated fire front at the same time after ignition, that is produced when the two fire lines are ignited simultaneously, with each allowed to influence the development of the other. (b) Simulated horizontal wind and vertical vorticity (10 m AGL) 10 minutes after ignition. The figure shows the formation of counter-rotating vortex pairs ahead of the fire line, which locally enhance the rate of forward progression of the fire line.

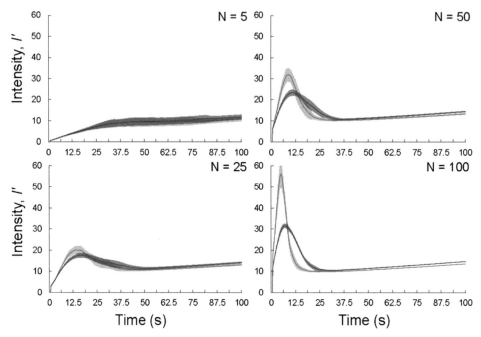

Figure 4.7 Spatial integral of normalized total fire line intensity (fire power) for spot fire coalescence over time for N = 5, 25, 50, and 100 spot fires. The lines show the mean value over 10 ensemble simulations and the shaded confidence bands show one standard deviation. Results from simulations without fire–atmosphere coupling are depicted in dark grey, and results from simulations including fire-atmosphere coupling are depicted in light grey.

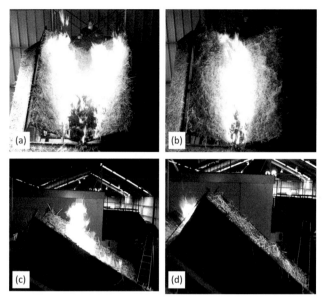

Figure 4.8 Fire behavior experiments in V-shaped canyons conducted at the Center for the Study of Forest Fires in Lousã, Portugal (University of Coimbra). Ignition was a point ignition in the centerline at the bottom of the canyon. (a) Plan view of fire spreading up a canyon inclined at 30° showing two distinct head fires; (b) Plan view of fire spread up a canyon inclined at 40°, showing a single head fire spreading up the canyon centerline; (c) Side view of the fire in the 30° canyon showing the flames rising vertically; (d) Side view of the fire in the 40° canyon showing the flames attached to the surface and erupting out of the top of the canyon.

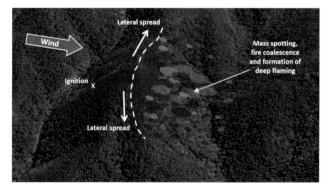

Figure 4.9 Schematic diagram illustrating a typical scenario involving VLS and associated downwind spotting, spot-fire coalescence, and formation of deep flaming. The fire has ignited mid-slope on a windward facing slope and initially spreads up the slope with the wind. As the fire encounters the ridge line (white dashed line), dynamic fire-atmosphere interactions drive the fire laterally (VLS). The regions of lateral spread act as an enhanced source of embers, which are deposited downwind as a dense ember attack.

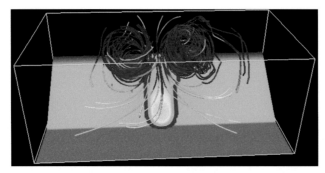

Figure 4.10 Coupled fire–atmosphere model (WRF-Fire) output with wind tendrils showing a counter-rotating pair of vortices at the ridge crest of a leeward slope. The fire has been ignited at the bottom of the leeward slope and has burnt back up the slope against the prevailing winds. The counter-rotating vortices are carrying the fire laterally across the top of the leeward slope. The vortex on the right is spinning in a clockwise direction (i.e. with negative vertical vorticity), while the one on the left is spinning in an anti-clockwise direction (i.e. with positive vertical vorticity).

Figure 4.11 Pyrocumulonimbus with anvil (*cumulonimbus incus flammagenitus*) over the 2013 Grampians fire, Victoria. Photo: Randall Bacon

Figure 4.12 Left: The 1994 South Canyon fire in Colorado (Butler 1998) – an example of a blow-up fire (Photo: SCFAIT (1994)). Right: The 2013 Forcett-Dunalley fire (Ndalila et al. 2020) – an example of an extreme wildfire. Photo: Rebecca White

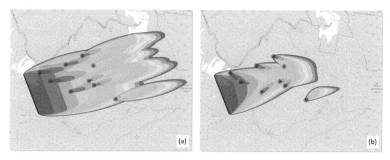

Figure 4.14 Fire spread simulations involving line and multiple point ignitions implemented in the Spark framework (Hilton et al. 2015; Miller et al. 2015): (a) Uncoupled simulation (i.e. without pyrogenic feedback) and (b) Simulation incorporating fire–atmosphere coupling using the pyrogenic potential model. In the uncoupled simulation, the fires spread independently of each other, while in the pyrogenic potential model simulation the fires interact with each other resulting in dynamic steering of the individual fires and a slowing of the overall progression of the fire complex.

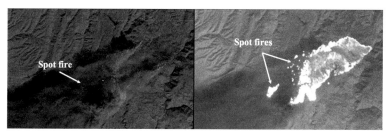

Figure 5.1 Evidence of wildfire propagation through fire spotting during the Camp Fire. This event erupted on November 8, 2018 near Sacramento, California. Images show active fire areas captured at 10:45 A.M. PST using the combination of the visible bands (2, 3, 4) and short-wave light sensor (left) and the thermal infrared sensor (right), which reveals more spot fires ignited ahead of the fire front.

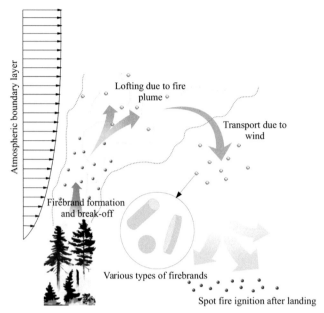

Figure 5.2 Schematic diagram illustrating the process of spot fire generation starting with firebrands formation, lofting into the atmosphere, downwind transport, and ignition of the fuel bed. Reproduced with permission from Tohidi et al. (2015)

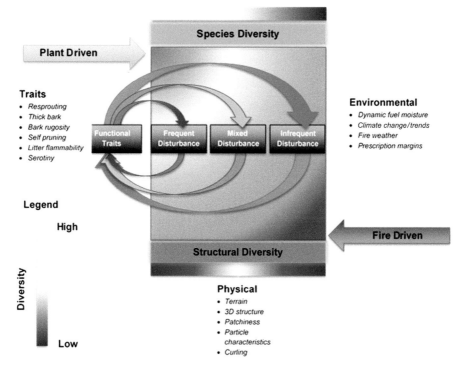

Figure 6.4 Synthetic model of plant–fire feedbacks. This conceptual model integrates the concepts of functional traits as a function of disturbance frequency (frequent, moderate, and rare). These functional traits are the evolutionary adaptations that plants develop to self-select for the type of fire and response that maintain their preferred disturbance regime. In this case, plant driven features are described as plants being the object and fire acting as the subject, where fire and plant response are a product of the plant adaptations. The reciprocal feedback to the plant driven model is a fire driven paradigm, where fire is the active object, generating larger fire effects that modify both structure and species diversity. These two endmembers are inextricably linked and feedback on each other over long temporal spans. Beyond functional traits, environmental and physical characteristics play a significant role in the ability for not only potential energy of fuels to be released, but also how, where, and when fire will affect ecosystems.

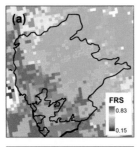

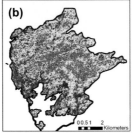

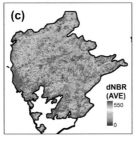

Figure 6.5 Patterns of mixed severity and disturbance example: North Rim Grand Canyon National Park, USA.

The north Rim of Grand Canyon National Park encompasses all three of the key disturbance type described in Figure 6.4. On lower peninsulas of the Kaibab Plateau, frequent fire in Ponderosa pine-gamble oak maintain functional traits as high lacunarity, thick bark for fire tolerance, and resprouting adaptations in deciduous shrubs after fire. At higher elevations, there are a mixture of mixed and rare disturbance regimes.

The Poplar fire occurred on the north Rim in 2003 in a matrix of mixed conifer and spruce-fire forest types burning 4,813 ha (Stoddard et al. 2020). The mixed severity Poplar fire burned through a broad range of fire resilience scores, an integrated metric of functional traits that quantifies the ability of the conifer system to maintain its current state after fire (a) (Stevens et al. 2020). Airborne laser scanning (ALS) derived structure classification from horizontal, vertical, and surface roughness metrics (Hoff et al. 2019) describes the structural variability resulting from the high severity patches (b), green and blue colors) in areas with low-to-moderate fire resilience scores. These sites represent state changes as a result of the high severity from mixed conifer and spruce-fire to resprouting aspen and New Mexico locust or grass land.

Comparison with the delta normalized burn ratio (dNBR) metric from the Monitoring Trends in Burn Severity program (MTBS; www.mtbs.gov (last accessed November 20, 2021)) with the ALS derived structure classes (c) shows predictable structural outcomes from heat flux, where fire resilient structure types that employ functional traits have porosity and lacunarity metrics that move convective heat efficiently.

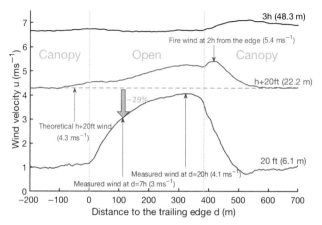

Figure 7.10 Horizontal wind profiles corresponding to Figure 7.9 at various heights. Bottom line: the wind velocity at sensor heights (20 ft), which exhibits a strong increase along the open until a distance of $20h$ to the trailing edge of the first forest block. Medium line: the wind velocity at $h + 20$ ft and in black the velocity at $3h$.

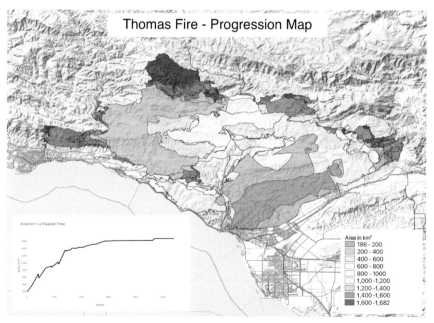

Figure 8.1 Fire progression map for the Thomas Fire, CA. Shading represents time over the period December 5–24, 2017. Note the complex evolution of the fire front boundary and the spread pattern, as it realigns with the changing wind and local topography. According to nearby RAWS the wind initially was gusting about 30 mph to the NE, with RH about 8%. Inset shows cumulative area (square kilometers) burned as a time series in hours from ignition. Data from GEOMACS

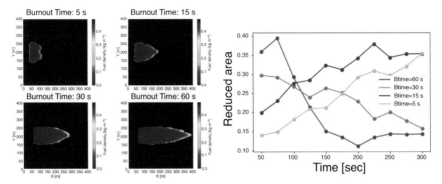

Figure 8.5 (Left) Fuel density for a homogenous grassland fire in a CFAM (QUIC-Fire) displayed at 300 seconds for four different burnout times, representing four fine fuel categories (30 seconds is a typical burnout time, or e-folding time scale for complete combustion of a model fuel cell element). (Right) reduced area as a function of time for the four burnout times; the two longer times show a decrease over time, while the two faster times show increasing reduced area. Reduced area decreases due to enhanced fingering with slower burnout times.

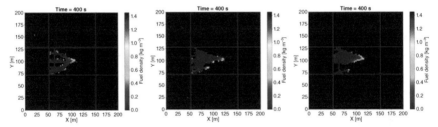

Figure 8.6 Fuel consumption at simulation end time (400 seconds). Fuel density remaining at the end of the simulation for the minimum (left), median (middle), and maximum (right) burned density for a low resolution, high wind, low moisture run.

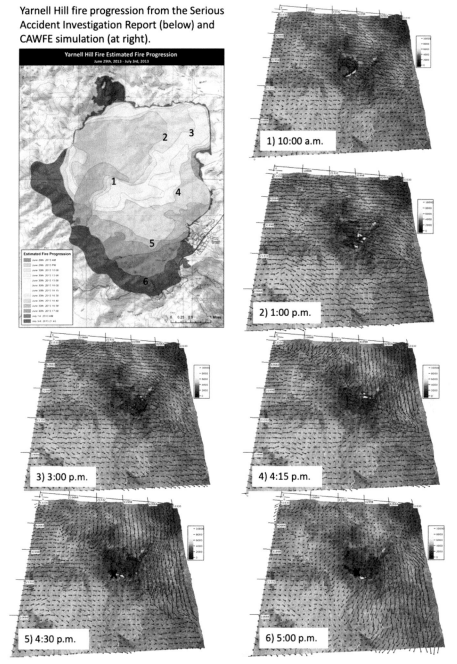

Figure 8.7 Yarnell fire progression from the Serious Accident Investigation Report (top left) and a sequence of times during a CAWFE simulation (numbered 1–6). Reprinted from Coen and Schroeder (2017)

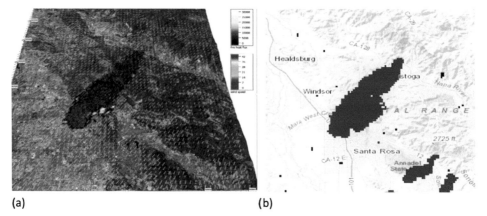

(a)   (b)

Figure 8.8 (a) CAWFE simulation of the Tubbs fire at 4:09 A.M. showing the downwind run of over 12 miles into the town of Santa Rosa in approximately 3 hours. The protrusion on the lower right side, orthogonal to the wind, occurred as the flank of the fire drew itself up a topographic feature. (b) Visible and Infrared Imaging Radiometer Suite (VIIRS) active fire detections at 3:09 A.M. October 9, 2017. Reprinted from Coen et al. (2018a)

5

Firebrands

ALI TOHIDI AND NIGEL BERKELEY KAYE

5.1 Introduction

One of the primary hazards to people and property during a wildfire is a set of spot fires generated by burning embers. There is increasing evidence that embers, alternatively known as firebrands, are a leading cause of home ignition in Wildland Urban Interface (WUI) fires (see Mell et al. 2010). Figure 5.1 shows evidence of this phenomenon captured by the Operational Land Imager (OLI) sensor on the Landsat 8 satellite during the Camp Fire; one of the most destructive wildfires in California, with more than 80 fatalities and over 18,000 structures lost. In fact, spot fire ignition of homes can occur before the arrival of the main fire front (see Caton et al. 2017a). The mechanisms of fire spotting are very well documented in the literature (Albini and Forest 1983; Fernandez-Pello 2009; Tohidi et al. 2015), as shown schematically in Figure 5.2. The four step process starts with firebrands breaking off from burning vegetation (Manzello et al. 2007) or structural elements (Suzuki et al. 2012). These combusting firebrands are then lofted into the fire plume by the thermal updraft generated by the fire heat. The firebrands are then transported downwind, and potentially ahead of the main fire front, by the ambient wind field. Finally, the firebrands land and, if they deposit on a receptive fuel source, a spot fire can be ignited.

The remainder of this chapter reviews the current state of knowledge on steps in the process of generating a spot fire. Section 5.2 examines firebrand formation and data on the size and shape of firebrands generated from wildfires. Section 5.3 reviews the various elements of firebrand lofting and transport modeling, including firebrand aerodynamics, the fire plume characteristics, models for the ambient wind field, model coupling, experimental results, and a discussion on the sensitivity of the model predictions to the inevitable uncertainty in the model input parameters. Finally, Section 5.4 highlights recent works on the physics of firebrand deposition and spot fire generation.

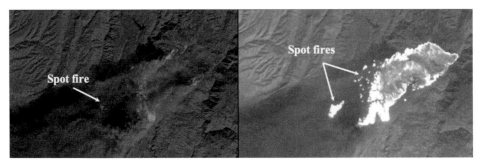

Figure 5.1 Evidence of wildfire propagation through fire spotting during the Camp Fire. This event erupted on November 8, 2018 near Sacramento, California. Images show active fire areas captured at 10:45 A.M. PST using the combination of the visible bands (2, 3, 4) and short-wave light sensor (left) and the thermal infrared sensor (right), which reveals more spot fires ignited ahead of the fire front. A black and white version of this figure will appear in some formats. For the color version, please refer to the plate section.

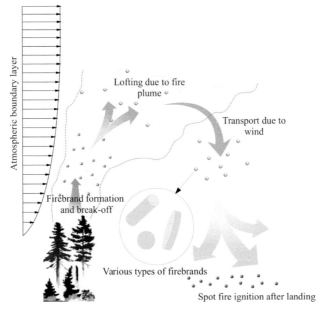

Figure 5.2 Schematic diagram illustrating the process of spot fire generation starting with firebrands formation, lofting into the atmosphere, downwind transport, and ignition of the fuel bed. Reproduced with permission from Tohidi et al. (2015) A black and white version of this figure will appear in some formats. For the color version, please refer to the plate section.

5.2 Firebrand Generation

Firebrands are generated when either loose debris such as fallen leaves, bark, or branches ignite or small portions of a larger combusting object, such as a burning building or tree, break off. The size and shape of firebrands formed by the ignition of forest floor debris (bark, twigs, and leaves) depend on the types of vegetation that form the litter. These firebrands, formed at ground level, can be blown out of the fire region ahead of the fire front, often at ground level. Firebrands formed by the break off from burning fuel from a structure or tree are more likely to be driven up by the fire plume and likely travel further than ground level firebrands (Tohidi and Kaye 2017c). The focus of this section is on firebrands formed by the thermal degradation of such combusting elements.

5.2.1 Firebrands from Structural Fires

There are numerous papers reporting on experimental studies that quantify the size and shape distribution of firebrands generated from various burn tests. For instance, Suzuki et al. (2012) ran a combustion test on a full-scale house and captured the generated firebrands in a set of 21 water pans. The authors report the distributions of firebrand mass and projected area for firebrands captured directly adjacent to the building and those collected 18 meters downwind. The results were compared to firebrand size distributions from tree burn tests and showed that the structural firebrands had larger projected surface area for a given mass. Other laboratory studies on firebrand generation from structural fires have also been conducted (e.g. Suzuki et al. 2014; Hedayati et al. 2019; Suzuki and Manzello 2019). Suzuki and Manzello (2019) investigated the influence of ambient wind speed on firebrand generation, and showed that there was a positive correlation between the average projected area of firebrands with the increase in wind speed. The impact of various cladding treatments on firebrand generation have also been studied in the literature; for more details see Suzuki and Manzello (2016).

5.2.2 Firebrands from Burning Vegetation

There are also numerous experimental results reported in the literature on firebrands generated from trees (Manzello et al. 2007, 2009; Mell et al. 2009). These studies are typically from single tree burns in a controlled laboratory environment. Similar to the aforementioned studies, the resulting firebrands are captured in water pans placed around the burning trees. As with the structural firebrands, these studies present data on the distribution of mass and projected area for the firebrands collected. Much of the data on tree burning tests are summarized in a detailed statistical analysis by Tohidi et al. (2015). This paper presented a statistical method for generating appropriately sized virtual firebrands for numerical flight simulations such that there is no notable statistical difference between the physical characteristics of the collected firebrands in laboratory studies and the virtual firebrands. Tohidi et al. (2015) also proposed a thermo-mechanical break-off model based on the results of Barr and Ezekoye (2013). Mechanical failure was assumed to occur when the net aerodynamic moment and moment due to the firebrand weight exceeds the capacity of the firebrand's connection to the larger burning vegetation. This simple, first order model indicated that the firebrand break-off in single tree burn experiments was controlled by the upward aerodynamic drag snapping firebrands off the tree rather than the net weight acting down. This was supported by observations during the experiments that showed firebrands being lifted up away from the burning trees.

Recently, Caton-Kerr et al. (2019) identified the formation mechanisms of firebrands from thermally-degraded cellular solids. This study establishes a framework through which the processes that lead to failure and eventually break off of the firebrands, namely firebrand generation, can be described. The mechanism of firebrand generation includes thermo-mechanical buckling of the material at micro-scale, physio-chemical interactions, and eventually failure due to external loading at macro-scale either in the form of traction or body forces. Although there is a lot of evidence to ascertain the chronological order of these processes, as shown in Figure 5.3 in a normal degradation condition, in extreme cases, that is wildfires, they may occur simultaneously or one process may dominate the others (Caton-Kerr et al. 2019). The following is a concise description of the firebrand formation mechanisms, as shown in Figure 5.3.

Persistent heat exposure to wooden material leads to the accumulation of water vapor and gases which build up a hydrostatic pressure inside the material and increase the internal pressure of the element. Due to physical properties of the wood (orthotropic characteristics), micro-cracks evolve in the tangential direction of the wood cross-section as the pressure increases. The formation of micro- and macro-cracks on the surface creates vents for the discharge of gases and water vapor, and induces a negative pore pressure which pulls the flame sheet toward the surface (Li et al. 2015). After this the material shrinks at micro-scale, exacerbating the effects of micro-cracks. Persistence of this cycle increases the thermal stresses and leads to buckling of the heat-exposed surface. As a result the location of cracks consolidates and propagates more through the material (Ventsel and Krauthammer 2001; Baroudi et al. 2017). At the same time, or maybe a short time after, pyrolysis changes the chemical composition of the material and primarily forms char. Consequently, the density and structural stiffness of the material decreases. The combined effects of the stress concentration and loss of structural stiffness help the cracks gain sufficient depth through the charred layer. This, effectively, reduces the load bearing capacity of the element and, once any external load is applied (e.g. wind induced drag), either the fibrous or brittle fracture occurs, which eventually leads to firebrands detaching from the fractured sections (Caton-Kerr et al. 2019).

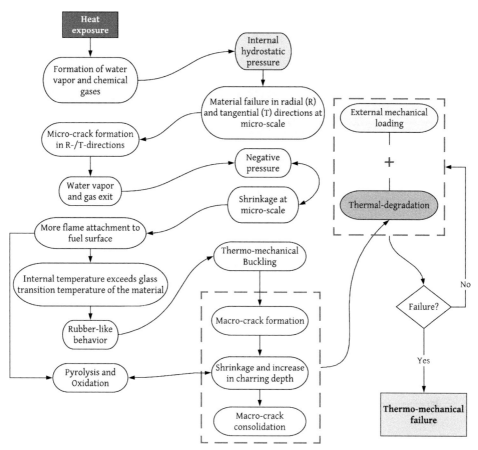

Figure 5.3 Flow-chart of the thermo-mechanical failure mechanisms in cylindrical wooden dowels under persistent external load and heat exposure. For more details refer to Caton-Kerr et al. (2019)

In order to determine the dominant fracture mode/regime in the failure of thermally-degraded structural elements under external loading, Caton-Kerr et al. (2019) also devised a set of three-point bending tests on heat-exposed cylindrical wooden dowels using a propane flame. Scaling analysis on the parameter space of their experiments suggest that the fracture mode in the thermally-degraded wooden dowels under external loading depends on the ratio of burning rate to the initial stiffness of the elements' cross-section, regardless of the type of wood and initial physical properties. These results are summarized in Figure 5.4, showing two dominant regimes. The first regime, that is, an approximately horizontal power-law fit, shows that the recoverable plastic strain (Π_2) is weakly affected by the burning rate parameter (Π_1). That is, the fracture is controlled by the stiffness of the material cross-section. However, the second regime (i.e. the very steep power-law fit in

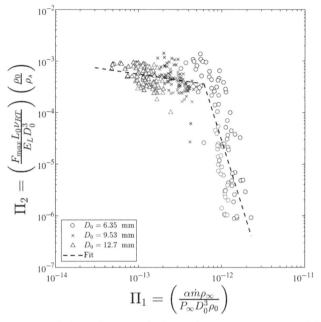

Figure 5.4 Results of dimensional analysis on the parameter space of Caton-Kerr et al. (2019)'s experiments. This result shows two distinct failure regimes that dominate the behavior of the thermally degraded wooden material under external loading. The first regime is fit by a weak power-law function and demonstrates the dominance of the mechanical fracture mechanisms, that is, weak/minor impact due to thermal degradation. And, the second regime fit by a steep power-law function shows the dominance of the brittle fracture due to combustion effects. For more details refer to Caton-Kerr et al. (2019)

Figure 5.4) is controlled by the burning rate which significantly affects the material's structural integrity (here, recoverable plastic strain, Π_2) and leads to brittle failure. For more details on the experiments and analysis of the results, see Caton-Kerr et al. (2019). Turning now to the firebrand shower modeling methodologies, Section 5.3 focuses on the transport models available in the literature.

5.3 Transport Modeling

The transport of firebrands through the atmosphere has many components that effect the eventual landing distribution of firebrands (spread). These components are studied either individually or coupled with other components through the literature. Early research on the transport phase of firebrand shower phenomenon was done by Tarifa et al. (1965, 1967) in which the drag force and burning rate of firebrands are identified as influential factors on the transport phase. Here, the components of transport phase are listed as,

1. Firebrand physical properties, which includes the density of the firebrands material and the size and shape of them. The firebrand's geometry and effective mass are typically described using probability distribution functions (Tohidi et al. 2015; Martin and Hillen 2016).
2. Combusting state as firebrands can continue to burn during flight. Combustion reactions cause the material to release heat and lose mass and, depending on the combustion stage (pyrolysis, oxidation, and charring), the shape of the firebrands may change as well. Further, either through mass loss, change in geometry or both a combusting firebrand shows a time-varying density behavior.
3. Aerodynamic properties of the firebrands determine their response to the ambient wind field. The landing distribution of firebrands is shown to have correlations with the aerodynamic coefficients (drag and lift) of the firebrands, which are themselves functions of the firebrand shape and flow angle, as well as the initial break-off (lofting) conditions (Tohidi and Kaye 2017a).
4. Fire plume model; The flow field generated by the large heat release from a wildfire is typically modeled as a turbulent plume. However, the type of the plume model will depend on the size and intensity of the fire and the ambient wind conditions.
5. Wind field characteristics; The ambient wind velocity drives the transport of firebrands ahead of the fire front. Atmospheric boundary layer models range in sophistication from simple steady-state uniform flow models to time varying turbulent flows resolved through intense computations.

As mentioned, each of the model components may be coupled with multiple other components. For instance, the combustion mechanism and rate depend on the air flow around the firebrand and its physical and chemical properties (Koo et al. 2010). In turn, combustion will alter the mass, size, and shape of the firebrands which leads to changes in their aerodynamic properties. The fire plume, also, can be influenced by the ambient wind field such that dominant winds may bend the fire plume over and decrease the lofting height of the firebrands.

The size, shape, and density of firebrands is dependent on the mechanics of firebrand formation, as discussed in Section 5.2. The combustion effects are complex and explicit inclusion of the thermo-mechanical processes into the transport simulation framework leads to intractable models. Hence, the combustion effects are, usually, accounted for in terms of empirical size and mass regression models that can be easily coupled with the transport equations (see, for example, Anthenien et al. 2006; Bhutia et al. 2010). The reader is referred to Sardoy et al. (2006, 2007, 2008), Kortas et al. (2009), Koo et al. (2010, 2012), and Albini et al. (2012) for more complex models and implementations of such effects into the transport phase. The remainder of this section reviews the state of knowledge on other components of firebrand transport such as aerodynamic characteristics of the firebrands (debris), fire plume, and wind field models. In addition, different coupling/modeling methodologies for the transport phase along with the available experimental data for model validation are discussed. Moreover, some discussions on the sensitivity of transport model outcomes to the initial conditions and model parameters are presented.

5.3.1 Firebrand Aerodynamics

Aerodynamic models for firebrand flight vary in complexity but are typically based on the debris flight equations initially formalized by Tachikawa (1983). The models were developed for estimating the flight distance, impact energy, and momentum of wind-borne debris in severe storms such as hurricanes and are based on Newtonian mechanics of rigid bodies. The model divided debris (firebrands in this context) into three geometric categories, namely compact, rod-like, and plate-like. For compact debris all three object dimensions are of a similar order of magnitude and the object is approximated as a sphere. Rod-like debris has one dimension that is much larger than the other two and is often approximated as a long cylinder. Plate-like debris has one short and two long dimensions, such as a piece of sheathing from a roof.

The simplest aerodynamic model for firebrands is the compact debris model. The model assumes that there are no aerodynamic lift forces (a lift force is an aerodynamic force acting normal to the flow velocity relative to the object) acting on the firebrand and that the only forces are drag and net weight. Therefore, these forces will only act on the object in two dimensions. That is, forces act only in the vertical plane parallel to the wind direction and the firebrand only moves within this plane. Applying Newton's second law to the firebrand leads to a set of coupled second order differential equations for the acceleration in the vertical and horizontal directions:

$$\frac{d^2x}{dt^2} = \frac{du}{dt} = \frac{\rho_{air}C_DA}{2m}(U-u)\sqrt{(U-u)^2+(W-w)^2} \qquad (5.1)$$

$$\frac{d^2z}{dt^2} = \frac{dw}{dt} = \frac{\rho_{air}C_DA}{2m}(W-w)\sqrt{(U-u)^2+(W-w)^2} - g, \qquad (5.2)$$

where x and z are the windward and vertical directions, u and w are the absolute firebrand velocities in the x and z directions, U and W are the x and z components of the wind field, ρ_{air} is the ambient air density, C_D is a drag coefficient, A is the cross-sectional area of the firebrand normal to the apparent wind direction (assumed constant), and m is the mass of the firebrand; see Kaye (2015) for a detailed derivation. For a firebrand in a steady uniform wind field, that is $W = 0$ and U is constant in time and space, the firebrand will eventually travel horizontally at the wind speed and vertically at its terminal velocity. For more complex wind fields, the equations must be solved numerically.

Equations (5.1) and (5.2) assume that firebrand flight is strictly two-dimensional. That is, there is no motion in the horizontal direction normal to the wind. It also assumes that there are no lift forces acting on the firebrand. This will only be true if the firebrand is perfectly spherical. Despite these limitations the compact debris flight model has been applied to non-compact debris such as pieces of gravel (Holmes 2004). The compact debris assumption has also been used implicitly to model the flight of disc shaped firebrands (Anthenien et al. 2006). This study aimed at establishing the maximum flight distance of disc shaped firebrands and assumed that the largest cross-sectional area of the firebrand was always normal to the apparent wind. This assumption ensured the maximum aerodynamic drag on the firebrand, as it was lofted and then blown down wind, envisioning that this would produce an upper limit on the potential flight distance of the firebrands.

The compact debris flight model eliminates any lift forces or net moments acting on the firebrand. However, this is very unlikely to occur in reality. For instance, lift forces can be generated by asymmetric flow separation over the object due to asymmetry in the firebrand shape or its rotation. There may also be aerodynamic moments applied to the firebrand that induce rotation and subsequent lift.

The rod-like and plate-like debris flight equations are an extension of the compact debris flight equations that account for these forces and moments. For each model there are six equations that model the aerodynamic forces and aerodynamic moments in each of the three coordinate directions. A full description of the model equations requires a coordinate transformation that tracks with the firebrand principal axes as they move during flight and is beyond the scope of this chapter. The reader is referred to Richards et al. (2002), Richards (2010, 2012), Grayson et al. (2012), and Tohidi and Kaye (2017a) for more details. The model requires detailed characterization of the aerodynamic drag, lift, and moment coefficients that are a function of the firebrand's shape and angle of attack (apparent wind direction). See Richards et al. (2002) and Richards (2010) for more details on these coefficients.

The flight of non-compact debris is highly sensitive to the variabilities in initial condition and to the characteristics of turbulence in the wind field (Tohidi and Kaye 2017a, 2017b). A simple experimental study in which model cylindrical firebrands were released under zero wind conditions showed that the resulting trajectory was extremely sensitive to the initial release angles (Tohidi and Kaye

2017a). The trajectories were so sensitive that it was not possible to model the path of an individual particle using the deterministic rod-like flight model. However, it was possible to predict the landing distribution of the model firebrands, statistically, by leveraging the same deterministic model but in a Monte Carlo fashion by changing the initial release angles randomly across many flight realizations. It is, therefore, possible to develop a statistical description of the flight trajectories of firebrands given a description of the wind field, and statistical distributions for the firebrand shape, and its initial release conditions.

5.3.2 Fire Plume Dynamics

Wildfires generate enormous amounts of thermal energy that heats and expands the air above the fire, making it buoyant. This buoyancy force generates vertical momentum that drives the air upward, creating a fire plume. Vertical updrafts from the fire plume can loft firebrands up into the atmosphere where they can then be transported downwind by the atmospheric boundary layer. The flow field within the fire plume is dependent on the size and strength of the fire and its interactions with the ambient wind field. Under calm conditions (zero wind), the fire plume will be vertical. For moderate wind speeds, the plume will be bent over by the wind with the degree of bending increasing with wind velocity magnitude. The plume's horizontal cross-section shape is determined by the shape of the area burning. For instance, for fires that are burning along a front with only burnt fuel behind the front (line fires), the fire plume would be planar in nature emanating from a long thin heat source. This section provides a brief overview of plume theory under no-wind conditions describing the role of source geometry and source fluxes of buoyancy and momentum on the plume development. The interaction of fire plumes with the ambient wind field is discussed in Section 5.3.4 on model coupling.

Far above a buoyancy source (fire in this case), a turbulent plume will rise vertically entraining surrounding air into the plume. As a result, the volume flux in the plume increases with height. The plume becomes less buoyant due to this mixing but the buoyancy flux (rate of transport of buoyancy) is conserved (provided the ambient fluid is not stratified). As ambient air is entrained into the plume the plume grows laterally with height. Dimensional analysis shows that, far from the source, the plume volume flux (Q), momentum flux (M), radius (b), and vertical velocity (W) scale like,

$$Q \sim F^{1/3}z^{5/3}, \quad M \sim F^{2/3}z^{4/3}, \quad b \sim z, \quad \text{and} \quad W \sim F^{1/3}z^{-1/3}, \tag{5.3}$$

where F is the plume buoyancy flux and z is the height above the source. Therefore, the plume grows linearly with height and the vertical velocity decreases with height.

The scaling relations have been confirmed both experimentally (Baines 1983) and theoretically (Morton et al. 1956); see, also, Lee and Chu (2003) for more details. The buoyancy flux of a fire is directly related to the fire heat release rate by

$$F = \frac{Pg}{\rho C_P T_\infty},$$ (5.4)

where P is the fire's heat release rate, g is gravitational acceleration, ρ is the reference density, C_P is the specific heat of air, and T_∞ is the absolute ambient temperature; see either Batchelor (1954) or Lee and Chu (2003) for a derivation of equation (5.4).

Nearer the source, the flow is more complex, with the standard plume equations no longer valid as the assumption of hydrostatic pressure does not hold true. Through that region, the fire plume is controlled by the balance of the source fluxes of buoyancy (related to heat release rate), volume, and momentum. The near ground plume dynamics are typically described using pool fire dynamic models. In general, due to the large temperature difference between the fire and surrounding ambient, there is a large buoyancy force that accelerates the fire plume away from the combusting fuel. The plume, therefore, contracts as it rises, leading to a neck before expanding as a pure plume far from the source.

There have been a number of approaches to modeling the near source adjustment of pool fires from the combustion (flame) zone up to the plume zone (Delichatsios 1987; Baum and McCaffrey 1989; Fay 2006), although the average statistics of the results relating to the velocity, temperature excess, and other parameters are similar in the near and far field. For instance, Baum and McCaffrey (1989), denoted as [BM89], presented a three layer model for a pool fire consisting of a flame zone, intermediate intermittent zone, and a plume zone. In the flame zone the fire plume accelerates but has a constant temperature. The intermittent zone has a constant velocity and decreasing temperature before transitioning to the plume zone through which both the temperature and velocity decrease with height. Fay (2006), denoted by [F06], presents a two zone model ("Combustion" and "Plume") with similar vertical scaling for the temperature and velocity, along with additional information on the flow rate and momentum flux in these two regions. Also, Delichatsios (1987), denoted by [D87], presents a three zone model, designated by the "Near Source," "Neck," and "Plume." In the aforementioned studies, the vertical scaling of the key parameters in a pool fire with height agree despite the different modeling approaches and the different parameter spaces considered. A summary of the pool fire scaling relations is given in Table 5.1, in which the numeric values correspond to the exponent, that is, (a), in the generic scaling equation written as (Plume Parameter) $\sim z^a$. Note that each study uses a slightly different method for scaling the variables in order to present their results in

Table 5.1 *Table of vertical distance power-law exponents for a range of pool fire parameters for different regions. References are to the origins of the region names and to the specific papers if not all papers reported a value*

	Region		
Parameter	Flame [BM89], Near Source [D87], Combustion [F06]	Intermittent [BM89], Neck [D87]	Plume
Velocity	1/2	0	–1/3
Temperature	0	–1	–5/3
Volume flux	3/2	1	5/3
Momentum flux	2	–	4/3

nondimensional form, which necessitates careful interpretation of the results for quantitative comparison(s).

5.3.3 Wind Field

While the fire plume lofts the firebrands above the fire, it is the ambient wind that drives them ahead of the fire front. As mentioned, various methods with different degrees of complexity have been used in firebrand transport studies. One method is to approximate the boundary layer with a uniform steady-state velocity profile, although this is a significant simplification compared to other atmospheric flow models. Nonetheless, this method allows for a first order estimation of the flight distance. In reality, the mean wind speed increases with height above the ground and is, usually, highly turbulent. The flow characteristics are controlled by prevailing weather systems in larger scales, such as atmospheric stability, and depend on the interactions of the boundary layer with the surface topography, land-use, and canopy height in cases of terrain covered with vegetation. Given this, a common approach is to model the time-averaged velocity field and derive the variations with height.

The logarithmic velocity profile, also known as log-law, is one of the well documented methods for approximating the vertical variation of the mean wind speed. The log-law formulation is

$$U(z) = \frac{U_*}{\kappa} \log\left(\frac{z}{z_0}\right), \tag{5.5}$$

in which

$$U_* = \sqrt{\frac{\tau}{\rho}} \tag{5.6}$$

is the skin friction velocity, τ is the surface wind-driven shear stress, ρ is the air density, $\kappa \approx 0.4$ is the Von-Karmen constant, and z_0 is the upstream surface roughness length. The values of U_* and z_0 can be calculated by curve fitting through measured wind speed data, although proper data measurement can be difficult for a given location.

A similar model with different formulation for approximating the vertical variation of the wind speed with height is the power-law velocity profile given by

$$U(z) = U_{ref} \left(\frac{z}{z_{ref}} \right)^{1/\alpha}, \tag{5.7}$$

where U_{ref} is a reference wind speed measured at a specified reference height z_{ref}. The reference height is the height above the ground at which the mean wind speed is measured at the nearest weather station. The value of $(1/\alpha)$ varies with the local terrain, with larger values associated with rougher terrains. For instance, the ASCE (1999) guide to wind tunnel testing of structures recommends values ranging from $1/\alpha = 0.15$ for open grass terrain to $1/\alpha = 0.33$ for densely packed forests. The advantage of this model for the mean wind field is that it can be established based on a single velocity measurement at a reference height and an assessment of the type of the local terrain.

The above models assume that the flow is fully developed over flat terrain. However, local topography will locally alter the nature of both the mean wind speed and its vertical variation. For example, as the wind flows up a ridge the flow accelerates. This acceleration has to be quantified in terms of speed up coefficients which are the ratios of the local wind speed to the wind speed at the same height upwind of the hill of obstruction; see Holmes (2001) for more details on modeling of wind flows for engineering applications.

A more complete model of the atmospheric boundary layer can be obtained through computational fluid dynamics (CFD) simulations that can account for detailed local topography and time-dependent fluctuations in the wind field. The disadvantage of this approach is that it is computationally intensive and, currently, not operationally feasible.

5.3.4 Model Coupling

In order to deterministically model the lofting and downwind transport of firebrands, or debris in general, the two components of the firebrand transport, namely the lofting and the downwind transport of the particles, need to be coupled. A variety of methods are presented in the literature, which range from completely decoupled models to fully coupled ones in which the plume and atmospheric boundary layer are resolved in a computational model as plumes in turbulent crossflows.

Decoupling the vertical lofting from the horizontal transport is a reasonable approximation of the flight in situations with very low wind speed, where the fire plume is predominantly vertical. In this case, the vertical velocity of the plume decreases with height and the maximum lofting height is where the vertical drag force balances the weight. This can be calculated analytically for compact debris. Hence, the wind-driven horizontal transport model can be initialized with firebrands that are released from this maximum rise height in order to estimate the downwind flight distance (Woycheese and Pagni 1999).

A slightly more complex coupling is to model the impact of the atmospheric boundary layer on the fire plume. Using this approach, the plume becomes bent over, due to the entrainment of air with horizontal momentum into the plume envelope. The bent over plume models vary from semi-empirical cases such as Briggs (1984) to modified versions of the plume's differential equations of Morton et al. (1956), to include the entrained horizontal momentum; see Hoult and Weil (1972) for example. Typically, these bent over plume models assume a uniform wind velocity profile, despite the straightforward incorporation of the atmospheric boundary layer model; see Tohidi and Kaye (2016) for the mathematical implementation and a more comprehensive review of the bent over plume models.

Once the bent over plume model is established the complete wind field can be approximated by using the bent over plume velocity field for firebrands within the plume and the wind model for firebrands outside the plume envelope; for example refer to Sardoy et al. (2008). This leads to a less computationally expensive model. However, such integral models for the plume behavior do not include turbulent fluctuations in the velocity field which may influence the flight dynamics of firebrands significantly.

A more computationally intensive method is to use a CFD model for resolving the time-varying flow field. This approach can be computationally intensive depending on the choice of turbulence model and the desired spatial resolution. In this approach, the Navier-Stokes equations of fluid motion (in this case air) are solved over the entire domain of interest including the fire plume to account for the velocity field. For modeling the turbulent fluctuations, the two most common approaches are to either solve for the Reynolds Averaged Navier-Stokes (RANS) with appropriate closure for the Reynolds stresses or to use a Large Eddy Simulation (LES) model. Using LES, the larger scales are resolved by solving for the time-varying velocity field and smaller scales are modeled by a sub-grid scale closure model. Although the LES models are more computationally expensive, they provide better understanding of the influence of the large- and small-scale flow structures (eddies) on the travel trajectory of the firebrands. Refer to Bhutia et al. (2010) and Tohidi and Kaye (2017b) as instances of using LES models in large and laboratory scale simulations, respectively, for firebrand transport modeling.

5.3.5 *Experimental Results*

There are numerous published experimental results on the size and shape of firebrands produced from the burning of various trees, structural elements, or even full-scale buildings, as discussed in Section 5.2. Also, many experimental studies on the accumulation of firebrands on or around various structures exist, as will be discussed in Section 5.4.1. However, very few controlled laboratory experiments can be found that measure the full firebrand transport trajectory.

The most relevant study is that of Tohidi and Kaye (2017c), who tracked the trajectories of model firebrands in a boundary layer wind tunnel. The tests were run using non-combusting rod-like model firebrands with different aspect ratios (i.e. 1, 4, 6), made from polyurethane. The velocity field from the interactions of the atmospheric boundary layer with the fire plume were modeled by releasing a turbulent air jet into a fully developed crossflow boundary layer generated by the large scale wind tunnel. Then, the model firebrands were released through the exit nozzle of the air jet with random initial release angles. The tests were run for each of the three firebrand shapes at three different reference wind speeds and three different lofting jet velocities. In order to quantify the variability in the flight trajectories, at least 200 firebrand releases were tracked (using the developed particle tracking algorithm) for each of these 27 test conditions. The experimental results showed that the firebrand shower phenomenon is stochastic and even under highly controlled laboratory conditions there is a large variability in the measured flight distance. For instance, in one set of tests for the model firebrands with aspect ratio 6, the standard deviation about the mean flight distance was 52% of the mean value. Sample plots of captured trajectories for the cubic firebrands are shown in Figure 5.5.

The primary goal of these experiments was to provide an experimental dataset that could be used to validate firebrand transport models. To that end, Tohidi and Kaye (2017b) compared the wind tunnel results with detailed flight models that coupled a Large Eddy Simulation (LES) CFD model of the wind fields, including the lofting jet, with the rod-like debris flight equations. Simulations were run for two of the nine experimental wind fields and all three model firebrand aspect ratios. The simulations provided a good estimate of the mean rise height and flight distance. However, the model overestimated the standard deviation of both these parameters. For more details refer to Tohidi and Kaye (2017a, 2017c).

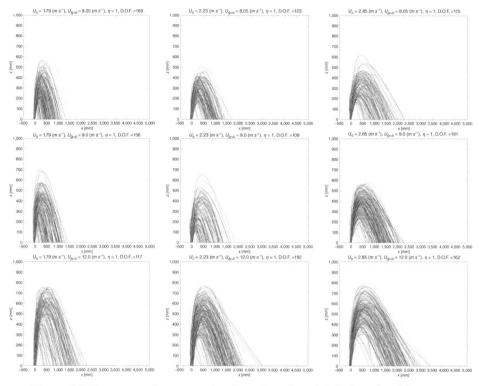

Figure 5.5 Shown are the captured trajectories of model firebrands with aspect ratio one released through the modeled jet in a crossflow velocity field. From left to right, the wind tunnel boundary layer reference velocities are 1.79, 2.23, and 2.85 m s^{-1}, and from top to bottom, the lofting jet centerline vertical velocities are 8.05, 9.0, and 12 m s^{-1}. Reproduced with permission from Tohidi and Kaye (2017c)

5.3.6 Probabilistic/Operational Models

A considerable amount of literature has been published on leveraging computational techniques, including intense CFD models, to simulate wildfire spread and particularly firebrand showers. As the general modeling methodology of this class of models are already explained in previous sections, the reader is referred to Sardoy et al. (2006, 2007, 2008), Kortas et al. (2009), Bhutia et al. (2010); Koo et al. (2010, 2012), and Tohidi and Kaye (2017b) for a few examples of computational models. In this section, the primary focus is on the probabilistic methods for operational models.

Although a number of intense (full-physics) computational studies (Kortas et al. 2009; Tohidi and Kaye 2017a, 2017b, 2017c) specifically attempt to model the stochastic behavior of firebrand transport and validate the numerical results with experimental data to further our understanding of firebrand shower phenomena, the developed models cannot be used operationally due to high computational complexities both in time and space. Hence, another class of models, which mostly work based on Huygen's principle (Pao and Varatharajulu 1976) and Rothermel's (1972) semi-empirical rate of spread (ROS) equation, have been developed; see Finney (1998) and Lautenberger (2013) for examples of operational models. Inherently the operational models, based on the ROS equation, do not account for the firebrand shower. However, to resolve this issue, recent works describe the motion (spread) of the fire as a composition of two main regimes, namely drifting and fluctuating (Kaur et al. 2016; Trucchia et al. 2019). The drifting regime refers to the fire spread that can be represented by the Rothermel (1972) ROS equation, while the fluctuating regime describes the fire spread due to stochastic factors such as turbulent heat flux and velocity field, firebrand shower, and the subsequent spot fires. In this method, the progression of fire in the drifting part would be calculated using the conventional ROS models and the progression that corresponds to the fluctuating part is calculated based on the probability density function (PDF) estimation of the ignition which depends on the physical characteristics of the firebrands, velocity field, etc.; see Kaur et al. (2016) for more details. Kaur et al. (2016) and Trucchia et al. (2019) assume that the drifting and fluctuating parts can be decoupled. Yet, this may only be valid in cases where the time-scale of the fire spread due to convection and radiation, as may be described by Rothermel (1972) ROS equation, is orders of magnitude different than the time-scale of the fire spread by spotting. For example, in grass fires the time scales are comparable and this assumption may propagate epistemic uncertainties into fire spread estimates.

Similarly, Martin and Hillen (2016) proposed a mathematical model for deriving the spotting distribution based on detailed physical processes and discussed effects of each process on the stochastic modeling approach. The main idea is similar to the approach of Kaur et al. (2016) and presents the fire spread as integro-PDE equations which employ kernels, based on physical processes, to describe long-distance firebrand dispersion. For example, the time evolution of the likelihood of fire at time t and at location $\mathbf{x}$, denoted by $u(t, \mathbf{x})$, within a gridded domain can be shown as

$$\frac{\partial u}{\partial t} = Du_{xx} + \int_{-\infty}^{\infty} \mathbb{S}(x - g)u(g, t)dg + c(u)u - \delta(u)u, \qquad (5.8)$$

where $\mathbb{S}(\mathbf{x})$ is the landing distribution throughout the domain, $c(u)$ accounts for combustion processes, $\delta(u)$ factors in the heat loss and the diffusion term accounts for the fire spread with $D_{\mathbf{x}}$, as the diffusion coefficient, being the rate of local fire spread. Martin and Hillen (2016), also, identify the key factor for estimating the spotting distribution as proper separation between the time-scale of the relevant processes such as wind-driven firebrand spread versus crawling spread of the fire line. One of the advantages of the probabilistic methods (Kaur et al. 2016; Martin and Hillen 2016; Trucchia et al. 2019) is that the firebrand shower (spotting) can be added as a postprocessing model to the current fire models.

Moreover, due to high degrees of spatial complexities and variabilities in the built environment and the stochastic nature of firebrand showers, as one of the main mechanisms of fire spread (Caton et al. 2017a), the probabilistic methods show great potential for capturing the behavior and providing more reliable estimates of fire spread in operational forecasts.

5.3.7 Sensitivity to Parameters

There are a large number of model input parameters for firebrand flight models including the firebrand size (mass and shape), the release conditions (location within the fire and launch angle), fire plume properties (heat release rate, size, turbulence properties), and wind field (mean and turbulent characteristics of the atmospheric boundary layer). For each of these sets of parameters there is uncertainty in the estimated quantities and natural variability in the underlying data. All of this uncertainty and variability can influence the resulting flight trajectories. In this section, we explore the sensitivity of the flight path to some of the key parameters.

As discussed in Section 5.2.1, there is a large degree of variability in the size and shape of firebrands produced by burning structures or vegetation. While there is a lot of experimental data on the probability distribution functions for various firebrand sources, much of the data is for isolated trees or structures under a limited range of wind conditions (for example, see Manzello et al. 2009; Suzuki and Manzello 2016). There is very little field data on the generated firebrands' size, shape, aspect ratio, and even number (flux) of the generated firebrands. Hence, proper and systematic collection of such field data is of paramount importance in understanding the firebrand shower phenomenon and, currently, a high research priority. See Thomas et al. (2017) for details on recent work on field data collection.

The launch location and initial release angle can significantly impact the flight distance. In Tohidi and Kaye (2017b) a series of simulations were run in which the location of the firebrand within the lofting vertical air current was varied by different amounts. They found that the more the release location was confined to the center of the lofting vertical air current the greater the flight distance. There is also significant variability in flight trajectory based on the initial angle of the release. A series of simple tests in which model firebrands were dropped in a quiescent ambient showed that the radial distance of the landing location from a point directly below the release point varied significantly for releases in which only the release angle was varied (Tohidi and Kaye 2017a). In fact, the coefficient of variation in the radial distance to the landing location was approximately 1. As a result, it is important to simulate a broad range of initial angles in order to quantify the range of possible landing locations.

Uncertainty in estimating characteristics of the fire will also impact the quality of any model to predict the flight distance. However, not all parameters have the same cascading effect. For example, the vertical lofting velocity in the fire plume scales on the cube root of the heat release rate (HRR); therefore, a 10% uncertainty in the HRR will only lead to a 3% uncertainty in the vertical lofting velocity. As a result, precise estimation of the HRR is not essential to effective flight trajectory modeling. On the other hand, turbulent fluctuations in both the lofting fire plume and the ambient wind field will significantly impact flight trajectory. For compact objects falling under gravity and driven horizontally by the wind, the mean flight distance is dependent on the root mean square velocity rather than the linear mean velocity and, therefore, the introduction of turbulent fluctuations into the wind field will increase the mean travel distance. See Karimpour and Kaye (2012) for more details. The impact of turbulence on the transport of noncompact debris is less clear. In the computation of Tohidi and Kaye (2017b) the average flight distance increased slightly when the time averaged (steady) wind field was used in the flight model. This is the opposite effect to that found by Karimpour and Kaye (2012), although this discrepancy may be due to the short simulation time for the wind field that was available for modeling and necessitates further research on this topic.

5.4 Ignition of Spot Fires

Upon landing firebrands can produce spot fires provided that they land on a susceptible fuel bed with a high enough temperature and thermal energy that can be transferred to the fuel bed. Herein, we first review ongoing research into firebrand accumulation with a specific focus on accumulation on structures. This is followed by a discussion on the conditions under which ignition will occur and available models.

5.4.1 Deposition and Accumulation

There are extensive research programs designed to assess the risk of combustion due to firebrand accumulation on structures. In fact, one motivation for the early tree burn experiments described above was to get calibration data for the design of firebrand generators that are now used to run structural accumulation tests. The

Figure 5.6 Image of the NIST firebrand generator (dragon) in a test of a roof section. Reproduced from Mell et al. (2010) with permission from CSIRO Publishing

firebrand generators, alternatively known as dragons, typically consist of a burn chamber, in which various wood chips are ignited, and a blower that lifts burning debris out of the chamber and blows it toward the target structure. An early firebrand dragon, built by a wildfire research team at the U.S. National Institute for Standards and Technology (NIST) (Manzello et al. 2008), had a single outlet and has been used extensively to investigate accumulation and ignition conditions for different structural components including roofing covers (Suzuki et al. 2016), wood decks (Manzello and Suzuki 2017), and in front of obstacles (Suzuki and Manzello 2017). See Figure 5.6 for an image of a dragon test of a roof panel. One drawback of this approach is that the tests do not attempt to model the wind flow around the entire building as well as the puffing generation mechanism as seen in the field. As such, they do not capture potential wind dead zones where firebrands might preferentially accumulate. To overcome the wind field interaction issue, the U.S. Institute for Business and Home Safety (IBHS) installed an array of multi-outlet dragons in their full scale wind tunnel to investigate the behavior of a full scale house when exposed to a model firebrand storm (Standohar-Alfano et al. 2017). In general, more insights from field data as well as controlled laboratory conditions are required to better understand the deposition mechanism in real scenarios.

5.4.2 Spot Fire Ignition

Ignition of spot fires induced by firebrands is likely the least understood phase of firebrand shower phenomena compared to the formation and transport phases. Spot fire ignition not only depends on the physical characteristics of the firebrands and their state upon landing – that is, flaming, glowing, smoldering, or charred – but also depends on the deposition morphology as well as the composition and physical properties of the fuel bed. In this section, the primary focus is on introducing the available ignition models for firebrands in the literature and review of research done on the ignition of wildland fuels as well as structural fuels.

5.4.2.1 Theoretical Models

As discussed in Section 5.4.1, the ignition risk due to firebrands is, also, dependent on the characteristics of deposition mechanisms on the fuel beds. There are very few studies on the theoretical aspects of the ignition due to firebrands, yet most of the theoretical models have considered the ignition of fuel beds by individual model firebrands in both hot reacting and hot non-reacting states. Recently, Manzello et al. (2020) provided an extensive review on the relevant theories. One of the main theories, inspired by ignition of gases by hot particles (Friedman 1963; Thomas 1965) and compact explosive material (Gol'dshleger et al. 1973), is based on the hot non-reacting model firebrands. In extending these theories to ignition of forest litter, it is suggested that the energy content of an individual model firebrand should be greater than a threshold for that particle to lead to the ignition of the fuel bed (Jones and Launder 1972; Jones 1993, 1995). Although this criterion does not account for the effect of particle size and fuel bed morphology, it leads to proper size–temperature relationships for ignition and has been adopted by the community (Rowntree and Stokes 1994; Babrauskas 2003; Hadden et al. 2011; Wang et al. 2015a, 2015b, 2017).

Based on Gol'dshleger et al. (1973) and later Hadden et al. (2011), the governing equations for a non-reactive hot particle, denoted by ($_p$), embedded in an unbounded medium with exothermic conversion properties in a condensed phase are,

$$\rho_p c_p \frac{\partial T_p}{\partial t} = \frac{\alpha_p}{V_p} \kappa \nabla T, \qquad 0 < x < r \tag{5.9}$$

$$\rho c \frac{\partial T}{\partial t} = \nabla \cdot (\kappa \nabla T) + \rho A \Delta H e^{-E/(RT)}, \qquad r \leq x < \infty, \tag{5.10}$$

where T is the temperature, t is time, ρ shows the density, c and c_p are the specific heat capacities, the surface area is shown by α, V denotes the volume, κ is the thermal conductivity of the material, A is the pre-exponential factor, E and R are the activation energy and universal gas constant, respectively, and ΔH is the heat of combustion. Equation (5.9) accounts for the temperature variations within the particle and Eq. (5.10) describes the temperature variations for the surrounding medium. The main objective of solving these equations is to find the critical particle size at which the ignition occurs. As stated, results of this theory qualitatively agree with experimental observations (Rowntree and Stokes 1994; Hadden et al. 2011) that as the model firebrand temperature decreases the critical size increases. For more details refer to Gol'dshleger et al. (1973) and Hadden et al. (2011).

Furthermore, a variety of numerical models (Zvyagils' kaya and Subbotin 1996; Grishin et al. 1998; Matvienko et al. 2018) as well as scale analysis (Yin et al. 2014), are developed to model the ignition of fuel beds by hot particles or firebrands and establish the ignition time, respectively. However, most of the models do not account for the porosity of the wildland fuels and the effect of various combinations in decomposition of firebrands on the fuel bed, as proposed in Hakes et al. (2019) and Salehizadeh (2019).

5.4.3 *Ignition of Different Fuel Beds*

Wildland fuels have a very complex morphology compared to solid beds. Generally, the wildland fuels consist of fine biomass that form a heterogenous porous material. The porosity level can vary a lot, from very small values to relatively large values for woody material such as twigs and branches. The moisture content level is another important factor in determining the characteristics of the ignition and can vary from moist to dry and live to dead. Other important factors are environmental conditions such as relative humidity, temperature, and wind conditions (Salehizadeh 2019). In addition, the ignition depends on the size, mass, and state of the firebrands upon landing, as discussed in Section 5.4.2. As a result, the ignition characteristics of wildland fuels is much more complex than solid fuel beds with homogenous morphology. To better understand this process several laboratory experiments have been conducted where wooden cylindrical dowels, in either a flaming or glowing state, were dropped on fuel beds of cellulose or sawdust to characterize the necessary conditions for a firebrand to ignite a flaming or smoldering fire (Fernandez-Pello et al. 2015; Urban et al. 2019). These experimental observations suggest that the dominant parameter in the firebrand induced ignition of such fuel beds is the moisture content of the fuel. For a comprehensive review of the studies on these fuel beds refer to Manzello et al. (2020).

During wildfires that occur at the wildland urban interface (WUI) areas, structures and buildings may ignite due to firebrand showers. The influential factors in this process not only encompass the parameters such as firebrand size, density, and state as well as the fuel bed type and environmental conditions, but also includes the construction type, geometric properties of the structural elements, physio-chemical properties of the components, density of vegetation cover around the structures, etc. Also, depending on the construction type the vulnerabilities of structural elements to firebrand induced ignition is different. In the literature, two construction types (i.e. US and Japan) are discussed; for a list of various structural components prone to ignition by firebrand showers refer to Caton et al. (2017b), Manzello (2014), and Manzello et al. (2020). Overall, due to more variability and complexities in the influential factors, the ignition of structural components by firebrands requires more research in order to further our understanding of the dominant parameters in the ignition phase of firebrand showers and its impact on the WUI areas.

References

Albini, FA, Alexander, ME, Cruz, MG (2012) A mathematical model for predicting the maximum potential spotting distance from a crown fire. *International Journal of Wildland Fire* **21**(5), 609–627.

Albini, FA, Forest, I (1983) Potential Spotting Distance from Wind-Driven Surface Fires. US Department of Agriculture, Forest Service, Intermountain Forest and Range Experiment Station, Ogden, UT.

Anthenien, RA, Tse, SD, Fernandez-Pello, AC (2006) On the trajectories of embers initially elevated or lofted by small scale ground fire plumes in high winds. *Fire Safety Journal* **41**(5), 349–363.

ASCE (1999) *Wind Tunnel Studies of Buildings and Structures*. Reston, VA: American Society of Civil Engineers.

Babrauskas, V (2003) *Ignition Handbook: Principles and Application to Fire Safety Engineering, Fire Investigation, Risk Management and Forensic Science*. Issaquah, WA: Fire Science Publishers.

Baines, WD (1983) A technique for the direct measurement of volume flux of a plume. *Journal of Fluid Mechanics* **132**, 247–256.

Baroudi, D, Ferrantelli, A, Li, KY, Hostikka, S (2017) A thermo-mechanical explanation for the topology of crack patterns observed on the surface of charred wood and particle fibreboard. *Combustion and Flame* **182**, 206–215.

Barr, BW, Ezekoye, OA (2013) Thermo-mechanical modeling of firebrand breakage on a fractal tree. *Proceedings of the Combustion Institute* **34**(2), 2649–2656.

Batchelor, GK (1954) Heat convection and buoyancy effects in fluids. *Quarterly Journal of the Royal Meteorological Society* **80**, 339–358.

Baum, HR, McCaffrey, BJ (1989) Fire induced flow field: Theory and experiment. *Fire Safety Science* **2**, 129–148.

Bhutia, S, Jenkins, M, Sun, R (2010) Comparison of firebrand propagation prediction by a plume model and a coupled fire/atmosphere large eddy simulator. *Journal of Advances in Modeling Earth Systems* **2**(1), 1–15.

Briggs, GA (1984) Plume rise and buoyancy effects. In: Haugen, DA, ed. *Atmospheric Science and Power Production*. Washington, DC: Department of Energy, pp. 327–366.

Caton, SE, Hakes, RSP, Gorham, DJ, Zhou, A, Gollner, MJ (2017a) Review of pathways for building fire spread in the wildland urban interface part I: Exposure conditions. *Fire Technology* **53**, 429–473.

Caton, SE, Hakes, RSP, Gorham, DJ, Zhou, A, Gollner, MJ (2017b) Review of pathways for building fire spread in the wildland urban interface part II: Response of components and systems and mitigation strategies in the United States. *Fire Technology* **53**, 475–515.

Caton-Kerr, SE, Tohidi, A, Gollner, MJ (2019) Firebrand generation from thermally-degraded cylindrical wooden dowels. *Frontiers in Mechanical Engineering* **5**.

Delichatsios, MA (1987) Air entrainment into buoyant jet flames and pool fires. *Combustion and Flame* **70**(1), 33–46.

Fay, JA (2006) Model of large pool fires. *Journal of Hazardous Materials* **136**(2), 219–232.

Fernandez-Pello, AC, Lautenberger, C, Rich, D, Zak, C, Urban, J, Hadden, R, Scott, S, Fereres, S (2015) Spot fire ignition of natural fuel beds by hot metal particles, embers, and sparks. *Combustion Science and Technology* **187**(1–2), 269–295.

Fernandez-Pello, C (2009) Modeling wildland fire propagation and spotting. In: *Fire Interdisciplinary Research on Ecosystem Services (FIRES), Seminar 3*. March 31–April 1, University of Manchester.

Finney, MA (1998) *FARSITE: Fire Area Simulator-Model Development and Evaluation*. Research Paper RMRS-RP-4, Revised 2004. Ogden, UT: US Department of Agriculture, Forest Service, Rocky Mountain Research Station.

Friedman, MH (1963) A correlation of impact sensitivities by means of the hot spot model. *Symposium (International) on Combustion* **9**(1), 294–302.

Gol'dshleger, UI, Pribytkova, KV, Barzykin, VV (1973) Ignition of a condensed explosive by a hot object of finite dimensions. *Combustion, Explosion, and Shock Waves* **9**(1), 99–102.

Grayson, M, Pang, WC, Schiff, S (2012) Three-dimensional probabilistic wind-borne debris trajectory model for building envelope impact risk assessment. *Journal of Wind Engineering and Industrial Aerodynamics* **102**(Mar), 22–35.

Grishin, AM, Dolgov, AA, Zima, VP, Kryuchkov, DA, Reino, VV, Subbotin, AN, Tsvyk, RS (1998) Ignition of a layer of combustible forest materials. *Combustion, Explosion and Shock Waves* **34**(6), 613–620.

Hadden, RM, Scott, S, Lautenberger, C, Fernandez-Pello, AC (2011) Ignition of combustible fuel beds by hot particles: an experimental and theoretical study. *Fire Technology* **47**(2), 341–355.

Hakes, RSP, Salehizadeh, H, Weston-Dawkes, MJ, Gollner, MJ (2019) Thermal characterization of firebrand piles. *Fire Safety Journal* **104**, 34–42.

Hedayati, F, Bahrani, B, Zhou, A, Quarles, SL, Gorham, DJ (2019) A framework to facilitate firebrand characterization. *Frontiers in Mechanical Engineering* **5**(Jul).

Holmes, JD (2001) *Wind Loading of Structures*. London: Spon Press.

Holmes, JD (2004) Trajectories of spheres in strong winds with application to wind-borne debris. *Journal of Wind Engineering and Industrial Aerodynamics* **92**(1), 9–22.

Hoult, DP, Weil, JC (1972) Turbulent plume in a laminar cross flow. *Atmospheric Environment (1967)* **6**(8), 513–531.

Jones, JC (1993) Predictive calculations of the effect of an accidental heat source on a bed of forest litter. *Journal of Fire Sciences* **11**(1), 80–86.

Jones, JC (1995) Improved calculations concerning the ignition of forest litter by hot particle ingress. *Journal of Fire Sciences* **13**(5), 350–356.

Jones, WP, Launder, BEi (1972) The prediction of laminarization with a two-equation model of turbulence. *International Journal of Heat and Mass Transfer* **15**(2), 301–314.

Karimpour, A, Kaye, NB (2012) On the stochastic nature of compact debris flight. *Journal of Wind Engineering and Industrial Aerodynamics* **100**(1), 77–90.

Kaur, I, Mentrelli, A, Bosseur, F, Filippi, J-B, Pagnini, G (2016) Turbulence and fire-spotting effects into wild-land fire simulators. *Communications in Nonlinear Science and Numerical Simulation*, **39**, 300–320.

Kaye, NB (2015) Solutions to the compact debris flight equations. *Journal of Wind Engineering and Industrial Aerodynamics* **138**, 69–76.

Koo, E, Linn, RR, Pagni, PJ, Edminster, CB (2012) Modelling firebrand transport in wildfires using HIGRAD/FIRETEC. *International Journal of Wildland Fire* **21**(4), 396–417.

Koo, E, Pagni, PJ, Weise, DR, Woycheese, JP (2010) Firebrands and spotting ignition in large-scale fires. *International Journal of Wildland Fire* **19**(7), 818–843.

Kortas, S, Mindykowski, P, Consalvi, JLL, Mhiri, H, Porterie, B (2009) Experimental validation of a numerical model for the transport of firebrands. *Fire Safety Journal* **44**(8), 1095–1102.

Lautenberger, C (2013) Wildland fire modeling with an Eulerian level set method and automated calibration. *Fire Safety Journal* **62**(Part C), 289–298.

Lee, JHW, Chu, VH (2003) *Turbulent Jets and Plumes: A Lagrangian Approach*. Boston, MA: Springer.

Li, K, Pau, DSW, Wang, J, Ji, J (2015) Modelling pyrolysis of charring materials: Determining flame heat flux using bench-scale experiments of medium density fibreboard (MDF). *Chemical Engineering Science* **123**, 39–48.

Manzello, SL (2014) Enabling the investigation of structure vulnerabilities to wind-driven firebrand showers in wildland-urban interface (WUI) fires. *Fire Safety Science* **11**, 83–96.

Manzello, SL, Maranghides, A, Mell, WE (2007) Firebrand generation from burning vegetation. *International Journal of Wildland Fire* **16**, 458–462.

Manzello, SL, Maranghides, A, Shields, JR, Mell, WE, Hayashi, Y, Nii, D (2009) Mass and size distribution of firebrands generated from burning Korean pine (Pinus koraiensis) trees. *Fire and Materials* **33**(1), 21–31.

Manzello, SL, Shields, JR, Cleary, TG, Maranghides, A, Mell, WE, Yang, JC, Hayashi, Y, Nii, D, Kurita, T (2008) On the development and characterization of a firebrand generator. *Fire Safety Journal* **43**(4), 258–268.

Manzello, SL, Suzuki, S (2017) Experimental investigation of wood decking assemblies exposed to firebrand showers. *Fire Safety Journal* **92**(Sept.), 122–131.

Manzello, SL, Suzuki, S, Gollner, MJ, Fernandez-Pello, AC (2020) Role of firebrand combustion in large outdoor fire spread. *Progress in Energy and Combustion Science*, **76**, 100801.

Martin, J, Hillen, T (2016) The spotting distribution of wildfires. *Applied Sciences* **6**(6), 177.

Matvienko, OV, Kasymov, DP, Filkov, AI, Daneyko, OI, Gorbatov, DA (2018) Simulation of fuel bed ignition by wildland firebrands. *International Journal of Wildland Fire* **27**(8), 550–561.

Mell, WE, Manzello, SL, Maranghides, A, Butry, D, Rehm, RG (2010) The wildland–urban interface fire problem: Current approaches and research needs. *International Journal of Wildland Fire* **19**, 238–251.

Mell, WE, Maranghides, A, McDermott, R, Manzello, SL (2009) Numerical simulation and experiments of burning douglas fir trees. *Combustion and Flame* **156**(10), 2023–2041.

Morton, BR, Taylor, GI, Turner, JS (1956) Turbulent gravitational convection from maintained and instantaneous sources. *Proceedings of the Royal Society of London A* **234**, 1–23.

Pao, Y-H, Varatharajulu, V (1976) Huygens' principle, radiation conditions, and integral formulas for the scattering of elastic waves. *The Journal of the Acoustical Society of America* **59**(6), 1361–1371.

Richards, PJ (2010) Steady aerodynamics of rod and plate type debris. In *17th Australian Fluid Mechanics Conference*. December 5–9, Auckland, New Zealand.

Richards, PJ (2012) Dispersion of windborne debris. *Journal of Wind Engineering and Industrial Aerodynamics* **104–106**, 594–602.

Richards, PJ, Williams, N, Laing, B, McCarty, M, Pond, M (2002) Numerical calculation of the three-dimensional motion of wind-borne debris. *Journal of Wind Engineering and Industrial Aerodynamics* **96**(10–11), 2188–2202.

Rothermel, Richard C. 1972. "A Mathematical Model for Predicting Fire Spread in Wildland Fuels." *Res. Pap. INT-115*. Ogden, UT: U.S. Department of Agriculture, Intermountain Forest and Range Experiment Station. 40 p. 115. https://www.fs.usda.gov/treesearch/pubs/32533.

Rowntree, GWG, Stokes, AD (1994) Fire ignition by aluminum particles of controlled size. *Journal of Electrical and Electronic Engineering* **14**, 117–123.

Salehizadeh, H (2019) *Critical Ignition Conditions of Structural Materials by Cylindrical Firebrands*. MSc thesis, University of Maryland.

Sardoy, N, Consalvi, JL, Kaiss, A, Fernandez-Pello, AC, Porterie, B (2008) Numerical study of ground-level distribution of firebrands generated by line fires. *Combustion and Flame* **154**(3), 478–488.

Sardoy, N, Consalvi, J-L, Porterie, B, Fernandez-Pello, A (2007) Modeling transport and combustion of firebrands from burning trees. *Combustion and Flame* **150**(3), 151–169.

Sardoy, N, Consalvi, J-L, Porterie, B, Kaiss, A (2006) Transport and combustion of Ponderosa Pine firebrands from isolated burning trees. In: *2006 1st International Symposium on Environment Identities and Mediterranean Area*, ISEIM, July 9–12, Corte-Ajaccio, France, pp. 6–11.

Standohar-Alfano, CD, Estes, H, Johnston, T, Morrison, MJ, Brown-Giammanco, TM (2017) Reducing losses from wind-related natural perils: Research at the IBHS research center. *Frontiers in Built Environment*, **3**(Feb.).

Suzuki, S, Brown, A, Manzello, SL, Suzuki, J, Hayashi, Y (2014) Firebrands generated from a full-scale structure burning under well-controlled laboratory conditions. *Fire Safety Journal* **63**(Jan.), 43–51.

Suzuki, S, Manzello, SL (2016) Firebrand production from building components fitted with siding treatments. *Fire Safety Journal* **80**(Feb.), 64–70.

Suzuki, S, Manzello, SL (2017) Experimental investigation of firebrand accumulation zones in front of obstacles. *Fire Safety Journal* **94**(Dec.), 1–7.

Suzuki, S, Manzello, SL (2019) Investigating effect of wind speeds on structural firebrand generation in laboratory scale experiments. *International Journal of Heat and Mass Transfer* **130**(Mar.), 135–140.

Suzuki, S, Manzello, SL, Lage, M, Laing, G (2012) Firebrand generation data obtained from a full-scale structure burn. *International Journal of Wildland Fire* **21**, 961–968.

Suzuki, S, Nii, D, Manzello, SL (2016) The performance of wood and tile roofing assemblies exposed to continuous firebrand assault. *Fire and Materials* **41**(1), 84–96.

Tachikawa, M (1983) Trajectories of flat plates in uniform flow with application to wind-generated missiles. *Journal of Wind Engineering and Industrial Aerodynamics* **14**(1–3), 443–453.

Tarifa, CS, Del Notario, PP, Moreno, FG, Villa, AR (1967) *Transport and Combustion of Firebrands*, final report of grants FG-SP-11 and FG-SP-146. US Department of Agriculture Forest Service.

Tarifa, CS, Notario, P, Moreno, FG (1965) On the flight paths and lifetimes of burning particles of wood. *Symposium (International) on Combustion* **10**(1), 1021–1037.

Thomas, JC, Mueller, EV, Santamaria, S, Gallagher, M, El Houssami, M, Filkov, A, Clark, K, Skowronski, N, Hadden, RM, Mell, W, Simeoni, A (2017) Investigation of firebrand generation from an experimental fire: Development of a reliable data collection methodology. *Fire Safety Journal* **91**, 864–871.

Thomas, PH (1965) A comparison of some hot spot theories. *Combustion and Flame* **9**(4), 369–372.

Tohidi, A, Kaye, NB (2016) Highly buoyant bent-over plumes in a boundary layer. *Atmospheric Environment* **131**, 97–114.

Tohidi, A, Kaye, NB (2017a) Aerodynamic characterization of rod-like debris with application to firebrand transport. *Journal of Wind Engineering and Industrial Aerodynamics* **168**, 297–311.

Tohidi, A, Kaye, NB (2017b) Stochastic modeling of firebrand shower scenarios. *Fire Safety Journal* **91**, 91–102.

Tohidi, A, Kaye, NB (2017c) Comprehensive wind tunnel experiments of lofting and downwind transport of non-combusting rod-like model firebrands during firebrand shower scenarios. *Fire Safety Journal* **90**, 95–111.

Tohidi, A, Kaye, N, Bridges, W (2015) Statistical description of firebrand size and shape distribution from coniferous trees for use in Metropolis Monte Carlo simulations of firebrand flight distance. *Fire Safety Journal* **77**, 21–35.

Trucchia, A, Egorova, V, Butenko, A, Kaur, I, Pagnini, G (2019) RandomFront 2.3: A physical parameterisation of fire spotting for operational fire spread models-implementation in WRF-SFIRE and response analysis with LS-Fire+. *Geoscientific Model Development* **12**(1), 69–87.

Urban, JL, Song, J, Santamaria, S, Fernandez-Pello, C (2019) Ignition of a spot smolder in a moist fuel bed by a firebrand. *Fire Safety Journal* **108**, 102833.

Ventsel, E, Krauthammer, T (2001) *Thin Plates and Shells: Theory, Analysis, and Applications*. New York: Marcel Decker.

Wang, S, Chen, H, Liu, N (2015a) Ignition of expandable polystyrene foam by a hot particle: An experimental and numerical study. *Journal of Hazardous Materials* **283**, 536–543.

Wang, S, Huang, X, Chen, H, Liu, N (2017) Interaction between flaming and smouldering in hot-particle ignition of forest fuels and effects of moisture and wind. *International Journal of Wildland Fire* **26**(1), 71–81.

Wang, S, Huang, X, Chen, H, Liu, N, Rein, G (2015b) Ignition of low-density expandable polystyrene foam by a hot particle. *Combustion and Flame* **162**(11), 4112–4118.

Woycheese, JP, Pagni, PJ (1999) Combustion models for wooden brands. In: Third International Conference on Fire Research and Engineering. Society of Fire Protection Engineers (SFPE), National Institute of Standards and Technology (NIST) and International Association of Fire Safety Science (IAFSS), Chicago, IL, pp. 53–71.

Yin, P, Liu, N, Chen, H, Lozano, JS, Shan, Y (2014) New correlation between ignition time and moisture content for pine needles attacked by firebrands. *Fire Technology* **50**(1), 79–91.

Zvyagils' kaya, AI, Subbotin, AN (1996) Influence of moisture content and heat and mass exchange with the surrounding medium on the critical conditions of initiation of surface fire. *Combustion, Explosion and Shock Waves* **32**(5), 558–564.

6

Re-envisioning Fire and Vegetation Feedbacks

ERIC ROWELL, SUSAN PRICHARD, J. MORGAN VARNER,
AND TIMOTHY M. SHEARMAN

6.1 Introduction

The discipline of characterizing wildland fuels originated with early models of crown fire (Byram 1959; Van Wagner 1975) and surface fire spread (Rothermel 1972). Although fire behavior modeling promoted the advancement of operational fire decision support, the models greatly simplified fuel inputs for the purposes of predicting wildland fire behavior to ensure firefighter safety. Contemporary fuel characterization is still often bifurcated into separate disciplines of fire behavior, using stylized fuel models along with canopy characteristics, and fire effects, which requires more detailed characterization of available fuels. The consequence of this division is a current lack of intersection among these models with fire behavior and two-way feedbacks that result from the three-dimensional arrangements in fuels, plant physiology, and evolutionary responses to regulating the fire environment. The focus of this chapter is to review the emerging discipline of fire–vegetation feedbacks for their inclusion in next generation fire behavior and effects models. We propose a re-envisioning of fuels from a vegetation perspective and research priorities for improved understanding of complex fire–vegetation feedbacks.

6.1.1 Background

Fire and vegetation dynamics are intrinsically a two-way feedback. How plants survive lethal fires or reinvade burned landscapes is a topic with wide-ranging implications from forest management, ecosystem productivity, nutrient cycling, and vegetation responses to climate change (Agee 1993; Keane 2015; McLauchlan et al. 2020). Fire is widely recognized as a fundamental ecosystem process in many terrestrial ecosystems (Agee 1993; Bond and Keeley 2005; Krawchuk et al. 2009; McLauchlan et al. 2020). The coevolution of plant traits that enable survival and persistence in fire-adapted ecosystems is increasingly acknowledged (Keeley and Zedler 1998; Varner et al. 2016; Stevens et al. 2020). However, early studies on fire–vegetation dynamics tended to focus on how fire impacts individual plants, plant populations, and communities (e.g. Agee 1993; Grimm 1994) and not the two-way feedbacks that exist, including plant adaptations to fire and contribution to fire behavior and effects. How plants modify fire through their structure, chemistry, moisture, phenology, and contribution to combustible biomass pools over time is less well recognized (Hoffmann et al. 2003, 2012; Pausas and Bradstock 2007; Pausas and Paula 2012; Varner et al. 2016). For their part, plants provide combustible biomass via their senesced leaves and woody branches and cones (dead fuels) and living leaves and fine branches (live fuels).

Until recently, characterization of live and dead vegetation (wildland fuels) has been motivated by required inputs surface fire spread modeling (Rothermel 1972), basic crown fire prediction (Byram 1959, Van Wagner 1975), and operational models of fuel consumption (Reinhardt et al. 1997; Prichard et al. 2017). For surface fire behavior modeling, fuel inputs are often simplified into associations via stylized fuel models that fit a simplified set of archetypes, from an original 13 fuel models (Anderson 1982) and 40 standard fuel models (Scott and Burgan 2005). Physical fuel properties, such as heat content or heat of combustion (kJ m^{-2}) and surface area:volume ratio (units of m^2/m^3 hence m^{-1}) by fuel type (e.g. shrub, herb, fine wood by particle size, and litter) are assigned within broad and homogenous representations of wildland fuels, and input fuel moistures are applied uniformly across modeled fire events.

Detailed fuel characterization that accounts for live and dead canopy and surface fuels is needed for fire effects modeling, including smoke production and tree mortality. Two of the most commonly used fuel consumption models, the First Order Fire Effects Model (FOFEM; Reinhardt et al. 1997) and Consume (Ottmar et al. 1992; Prichard et al. 2006), require estimates of mass per unit area of fuels within major fuel strata, including trees, shrubs, herbs, downed wood, and litter. The Fuel Characteristics Classification System (Ottmar et al. 2007) contains a modified version of the Rothermel fire spread model that allows for more detailed fuel characterization into fuel strata and categories. However, even within these more detailed fuel characterizations, fuels are summarized as one to two-dimensional representations of combustible biomass and are generally represented as single estimates of mass per unit area (kg ha^{-1}) of tree crowns, shrubs, herbaceous fuels, downed wood, litter, and duff at coarse spatial scales (Mg ha^{-1}). For example, fuel layers within LANDFIRE (Reeves et al. 2009; Rollins 2009) are widely used for fire behavior and effects modeling and provide estimates of combustible biomass in 30 meter raster grids.

Although still widely used today, metrics that simplify the structure and composition of surface and canopy fuels are incapable of capturing fire–vegetation dynamics that are the basis for prescribed burning and fire effects. For example, fire managers frequently employ strategies for prescribed burning that rely on fuel breaks that are comprised either of bare mineral soil or live vegetation with fuel moisture that present barriers to fire spread (e.g. managed fuel breaks or seasonal green grass; Agee et al. 2000; Agee and Skinner 2005). Trees and shrubs also modify wind and fire flow through sites and are important factors in predicting fire spread, duration, and effects (Parsons et al. 2011; Pimont et al. 2011). Although the structure and composition of wildland fuels has long been recognized as critical for understanding fire behavior and effects, models that accommodate realistic fuelbeds are still under development and evaluation and have not been adopted operationally.

Next-generation fire models, including the computational fluid dynamics model FIRETEC (Linn and Cunningham 2005) and the Wildland–Urban Interface Fire Dynamics Simulator (WFDS, Mell et al. 2007), have the capacity to embrace a broader spectrum of plant traits. These models can specify fuel composition and structure and fuel moisture in three dimensions (3D). However, because fine-scaled estimates of 3D fuel structure and composition are difficult to obtain, surface vegetation is often still represented as homogenous fuelbeds (constant mass per unit volume) in model simulations (e.g. Linn et al. 2013; Hoffman et al. 2015; Marshall et al. 2020). Recent advances in remote sensing technologies are allowing the development of three-dimensional characterization and mapping of wildland fuels. These advances will facilitate the evaluation of how wildland fire interacts with combustible biomass in three dimensional space, including the influence of fuel physical properties (e.g. bulk density, surface-area:volume ratio, heat content), condition (i.e. spatial configuration and temporal dynamics of fuel moisture), as well as the structural dimensions of where fuels are and gaps in fuel structure.

6.2 Fuels Re-envisioned: The Ecology of Fuels and Fire Environments

6.2.1 Redefining Wildland Fuels: The Need for 3D Characterization and Metrics

Fundamentally, wildland fuels are stored potential energy that can be released through pyrolysis and combustion. Plant flammability traits (Varner et al. 2015), physical environment (Waring and Running 1998), three-dimensional organization, and ambient weather/climate govern thresholds to burning and how efficiently fuels combust and contribute to fire spread (Byram 1959). Availability of this stored potential energy for combustion is directly tied to the capacity for fire to ignite and propagate from surface fuels that are typically complexes of deposited litter (needle/broadleaf), coarser fuels (branches, cones, and twigs), grasses, forbs, and low shrubs (Sandberg et al. 2007).

The structure and composition of wildland fuels dynamically change over time, dependent on plant traits and phenology, site ecology, and disturbance processes. Keane (2015) describes wildland fuel dynamics in the context of development, deposition, decomposition, and disturbance. As such, wildland fuels are defined not simply as combustible biomass but dynamic pools of live and dead vegetation, intrinsically tied to plant growth, succession, and disturbance processes. Biophysical setting influences fine- to coarse-scale patterns of vegetation cover and productivity, which subsequently defines the spatial and three-dimensional organization and amount of fuels (Rowell et al. 2020).

The ecology of vegetation, including how plants interact, grow and change over seasons and years, and in relationship to disturbance processes, have important implications for the structure and composition of wildland fuels. How plant species utilize and optimize water, light, and nutrients strongly define the 3D structure of wildland fuels. Examples of optimization in trees include inter- and intraspecific thinning based on competition for light and/or water, self-pruning for optimization of height and foliar biomass allocation (Mäkelä 1997), and plasticity of crown morphology in variable density forests (Delagrange et al. 2008; Vincent and Harja 2008). Additionally, many pine species shed older needles to populate the crown with young needles that have higher photosynthetic efficiency (Pallardy 2008). Optimization for survival and growth is not only ubiquitous in trees; understory plants also react to resource scarcity by, for example, having larger leaves in understories with limited light or providing greater hydrological conductance due to deep tap roots (Renninger et al. 2015).

6.2.2 Multi-scaled Metrics of Wildland Fuel Structure

Determining the physical properties of fuels and how they contribute to energy release is a key aspect of fire behavior and effects modeling. However, because fire burns as a contagious process, characterizing where fuels are absent in three-dimensional space is also critical to assessing thresholds to ignition and fire spread. As represented by the fire triangle, fires require fuel, oxygen, and heat to burn. Fine gaps in fuel structure concentrate air flow and influence oxygen availability, thus potentially amplifying fire intensity and spread (Hiers et al. 2009). At fine spatial scales, the term *porosity* (Anderson 1969) or packing ratio (Rothermel 1972) describe the compactness of a fuelbed and the availability of oxygen to infuse fuels to combust efficiently. For example, with optimum packing ratio, fuelbeds have enough pores (open space) to provide oxygen for combustion but are close enough to each other to transfer heat and thermally decompose through pyrolysis. Highly compact fuels, such as organic soils, lack air space and available oxygen, which produces an inefficient fire environment dominated by smoldering combustion. In contrast, gaps in fuel structure can render fuels less conducive to

Figure 6.1 Terrestrial laser scanning (TLS) derived profiles of mixed forests, depicting large gaps (lacuna) in canopy structure that affect fluid flow from the greatest impedance to surface fuels in (a) a spruce-fir matrix, (b) mature aspen with emergent spruce understory that has gaps for fluid flow around the conifers and a diffuse overstory that is conducive to convection and advection of air through the canopy, and (c) a frequently burned longleaf pine stand that optimizes insolation and fluid flow of air from a diffuse and sparse canopy that has substantial gaps between the crown and flammable understory.

fire spread during calm weather or, alternatively, facilitate increased air and fire flow under strong winds and/or extreme fire behavior.

The term *lacunarity* has been introduced to represent the sum of units of canopy gaps (Figure 6.1) that is observed as a result of fractal geometry (Mandelbrot 1983; Zeide 1993). However, there has been little work that has defined the porous nature of fuels as a continuum from surface to canopy. A conceptual framework that describes the interplay of porosity and lacunarity is presented in Figure 6.2. Within this framework, we evaluate the three-dimensional distribution of wildland fuels at two distinct scales, the scale of lacunarity, represented by white space, and finer-scaled porosity, represented as a gradient of dark gray (low porosity) to light gray (high porosity).

In most wildland fuelbeds, the bulk of available surface fuels are represented by fine-grained lattices of forest litter, grass, and fine wood. How densely packed the lattice is sets the extent to which it impedes or allows for entrainment of air through the porous medium. Surface fuels that are typically an amalgamation of fine particles when loosely packed (high porosity) contain gaps and access for oxygen to entrain within the fuelbed, increasing the efficiency of combustion. Long-needled pine litter or complexes of pine and broadleaf litter often have the optimum packing ratio to support rapid fire spread (Varner et al. 2015). Larger objects such as pinecones and coarse wood may be part of these fuel complexes and partly consumed during flaming combustion. However, due to lower surface-area to volume ratios, these fuel elements require preheating and desiccation to fully combust, and generally continue smoldering after the passage of the flaming front. There are many aspects to releasing the potential energy of wildland

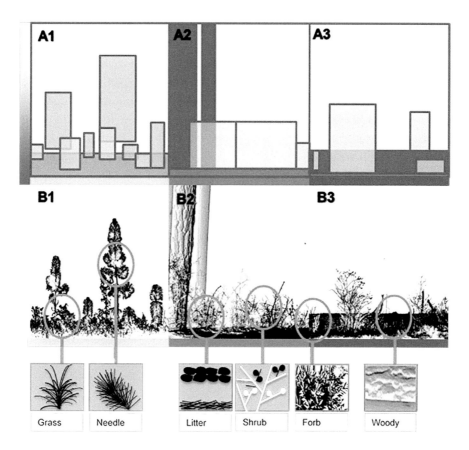

Figure 6.2 Fuelbeds can be described as a lattice or matrix of combustible features and obstructions that respond to moisture dynamics and carry fire. The interplay of the concept of lacunarity and porosity (A1–3) describes the bulk movement of air and the fine scale moderator of fluid flow, both acting as a flow diffuser due to the organization, shape, and distribution of fuel particles. The conceptual framework of porosity is depicted for a series of fuel types, ranging from a regenerating longleaf site that is comprised of diffuse fuel elements (B1), a western pine site composed of tree litter that are veneers of fuel that are comprised of fuel particles which act as plates or cylinders with shrub lattices (B2), and a post high severity site that is an organization of forbs, and obstructions such as dead and downed trees (B3).

fuelbeds, but the integration of all of these variables is highly dependent on vegetation, past disturbances, fire weather, and fuel moisture.

6.2.3 Consequences of 3D Fuel Structure for Fire/Atmosphere Interactions

Because fuel stratigraphy and three-dimensional structure are critical to fire spread and wind flow, the need for a porosity metric that integrates from surface to canopy fuels is necessary for next-generation fire modeling. In computational fluid dynamics models of fire behavior, combustion is represented as a fluid flow through a porous medium, influenced by interactions with a turbulent atmosphere. As such, large obstacles in the path of air movement, such as trees or shrubs, influence the flow of air, much like a boulder in the middle of a river.

During wind-driven fire spread, vortices form on the leeward side of large objects such as tree boles and increase the temperature and duration of the leeward flames (Gutsell and Johnson 1996). When wind velocity is high, trees with larger stems can amplify this effect. Large logs (i.e. coarse woody debris) on the forest floor also can be a direct impediment to the spread of surface fires, disrupting the continuity of fine surface fuels (Hermann 1993) and influencing how wind and fire spread through burn units.

Tree crowns, in isolation or as part of a canopy layer, create drag on wind flows which influence fire behavior both through lacunarity (e.g. crown base height) and porosity (e.g. crown bulk density). Although traditionally assumed as a continuous and homogeneous layer of uniform height in models of fire spread (Van Wagner 1977; Rothermel 1991), newer models have recognized the reality that tree canopies can be a heterogeneous layer of differing heights and structures, including porous and nonporous elements (Parsons et al. 2011; Pimont et al. 2011). Dense patches of regenerating trees and shrubs can also influence fire and wind flow. Unlike solid objects, shrub and tree canopies are porous, modifying but not completely impeding air flow. Parsons et al. (2011) compared modeled fire behavior between homogeneous and heterogeneous tree crowns in both low and high intensity surface fires. They found that, in the low intensity simulations, there was nearly 80 percent crown consumption in the homogenous crown canopy, while the heterogeneous canopy did not ignite (Parsons et al. 2011). In the high intensity surface fire simulation, the heterogeneous crowns had more complex fire behavior, igniting slower but having higher net radiative heat transfer (Parsons et al. 2011). These modeling examples illustrate the complexity of interactions between tree crowns and fire behavior and how simplifying assumptions are poor approximations of real fire behavior.

Tree crown structure and foliage can have an interactive effect on fire behavior. It is largely believed that self-pruning in pines is a fire adaptive trait that reduces ladder fuels and prevents surface fires from crowning. Further, having deciduous foliage may affect fire behavior in colder climates. For example, Wang (2002) found higher fire severity in boreal stands with a higher basal area of coniferous species compared to those with more deciduous species. The conifer dominated stands had high consumption of both litter and duff. While litter flammability likely played a role, one major contributor to the high severity was the fact that the coniferous crowns (*Abies* spp., and *Picea* spp.) prevented snow from accumulating on the ground near the boles, resulting in lower duff moisture compared to the deciduous species (*Populus tremuloides*), which were still in the leafless state.

6.2.4 Flammability of Live and Dead Vegetation

The three-dimensional distribution of fuels is only a partial influence of the fire combustion environment. How "available" fuels are for ignition and spread is also determined by their status (live/dead), moisture content, and physical properties, including the surface-area to volume ratio (SA:V) and particle size (e.g. of leaves, fine branches, downed wood, and litter), heat content and chemical composition, and bulk density (mass per unit air volume-time).

Plants modify the combustion environment via their living or dead parts in local neighborhoods of influence (Williamson and Black 1981). Based on flammability experiments in dried needles and leaves, there is a large body of evidence that species exhibit wide variation in their ignitability (Anderson 1970), intensity and duration of burning (sustainability and combustibility), extinction rates, and the proportion that is consumed (consumability; Varner et al. 2015). From a fire behavior modeling perspective, intensity depends on the shape of a fuel particle; pine needles and leaves, for instance, are typically distinguished as either a cylinder or a flat plate denoted by their SA:V ratio or particle size to determine fire reaction intensity (Burgan and Rothermel 1984).

Senesced large leaves burn with greater intensity, particularly those that are thinner in cross-section and with high perimeter to area ratio (P:A), and are prone to curling when dried (Engber and Varner 2012; de Magalhães and Schwilk 2012; Pausas 2015; Varner et al. 2015). Numerous studies have evaluated the respective flammability of two dominant temperate genera: *Pinus* and *Quercus*. This work demonstrates that morphology, chemical content, and moisture content of leaves produces significant differences in flammability (Fonda 2001; Kane et al. 2008; Engber and Varner 2012). Beyond how they burn, species differ in how rapidly they gain or lose moisture, making their fuel more readily available for combustion while other species remain too wet to combust (Kreye et al. 2018). Living foliage resists ignition due to their high moisture content and their position above the flaming zone. Experiments have demonstrated an expected sill in temperature as heat is absorbed by moisture within the live fuels. These live fuels experience a longer duration of flames in leaves above flames, where heat and oxygen movement are affected by wake effects due to leaves in the lower strata of shrubs (Pickett et al. 2009, 2010).

Living and dead vegetation on the soil surface are diverse fuel elements and interact with flows in diverse ways. Herbaceous understory components of the fuelbed are characterized as a more porous medium than the forest floor; however, these fuelbeds are comprised of a matrix of live and dead material that require variations of fire intensity to engage in combustion. Bunchgrasses and many other herbaceous species are characteristically porous, enabling rapid oxygen infusion and intense fires for their diminutive stature. For example, individual clumps of the understory dominant bunchgrass *Aristida stricta* (syn. *A. beyrichiana*) are readily ignitable and can generate flames 3 meters or more long (Fill et al. 2016). Other grasses and some forbs are capable of similar flammability. Shrubs are components of understory and midstory fuelbeds, interacting in several ways with fire. Shrubs intercept overstory litter, suspending senesced foliage as draped fuel that magnifies shrub ignition and intensity. Shrubs also cast shade to litter and herbaceous fuels beneath them and serve as a drag to convective flow that impedes surface fire intensity. In surface fire spread, shrubs ignite and burn as small "crown fires" likely following similar rules (height of crown fuel, foliar moisture content, and bulk density) that we traditionally assign to overstory canopy fuels (Van Wagner 1977). Downed wood, across the spectrum from fine branches to coarse logs, resist ignition due to their bulk and low surface-area:volume ratio and similarly impede flow. Once ignited, they generate residual heat that conducts heat to underlying organic and mineral soil where living plant roots can be injured, thus generating open spaces over time.

6.2.5 Fuel Moisture

Live and dead fuel moistures are critical to determining whether fuels ignite and how much fuel is available to burn over time and space. As such, the configuration of live and dead fuel moisture contributes to the four-dimensional structure of where wildland fuels are rapidly available to burn and where they are not. Living foliage contains around 50% to over 300% of its dry weight. As such, fuel moisture represents a significant heat sink, as water must be vaporized from live fuels before they can ignite. Live fuel moistures are primarily controlled by ecophysiological processes, such as phenology, transpiration, evaporation, and soil water, which differ among taxa and across regional climate (Jolly et al. 2014). In plants with complex architecture (i.e. shrubs and trees), moisture content can vary across foliage and fine branch architecture and also can be highly dynamic across time, with marked differences between daytime moisture content, deficits during intense solar radiation, and recovery following periods of shade or darkness. Moisture dynamics of dead fuels, including fine downed wood and litter, are strongly tied to vegetation structure, climate (because their time lag is so slow), soil moisture, and ambient weather (solar radiation, water vapor deficit) and the physical properties of fuels (e.g. size, density, surface area) (Fosberg et al. 1970; Viney 1991; Keane 2015).

Spatiotemporal dynamics of live and dead fuel moisture are often key determinants of whether fuels are available to burn, either promoting or inhibiting fire spread. For example, across ecosystems with grass-dominated fuelbeds, spring green-up, with more than 200% moisture content, is generally considered a barrier to fire spread (Gruell et al. 1986). However, the same sites can readily support fire spread following grass senescence and curing (less than 30–40% moisture). Similarly, the "spring dip" in conifer species is characterized as a drop in fuel moisture prior to bud burst (Little 1970; Chrosciewicz 1986). During this short period of time prior to the growing season, subboreal and boreal mixed conifer forests can be particularly conductive to crown fires (Jolly et al. 2014). Live and dead fuel moisture is closely tied to plant physiology and phenology. The inherent moisture content of attached living foliage undergoes change over time, with very high moisture contents at leaf expansion that declines over its lifespan. Foliage and branches have structures that while living prevent water loss (for foliage, a waxy cuticle, extra-stomatal structures, and photosynthetic pathways; for branches and boles, thick bark; Pallardy 2008). The phenophase (observable stages in the lifecycle of a plant) also effect the live fuel moisture of plants throughout the growing season via flowering and fruit development that in some species increase the probability of combustion in the fruiting phenophase (Emery et al. 2020). When senesced, these water-retaining traits degrade through weathering, enabling a dynamic of adsorption and absorption of water with wetting and desorption with drying (Fosberg et al. 1970). Decomposition alters moisture dynamics further, accelerating water gain and loss over time (Kreye et al. 2018). Seasonal to even diel differences in fuel moisture and the corresponding availability of fuels to burn are well known among prescribed fire practitioners, and research supports this (Banwell et al. 2013), but these fundamentals are not explicitly represented in predictive fire behavior, fuel consumption, and smoke models.

6.3 Two-Way Feedbacks between Fire and Vegetation

Vegetation and fire are inextricably linked, with each influencing the other in space and time. Substantial research has been undertaken on how fires affect vegetation and individual plants. Although traditional research has emphasized fire impacts to vegetation in terms of mortality and as a change agent, plant adaptions to fire – including pruning, vigorous sprouting, and regeneration – all convey a more evolved interaction with fire in fire-adapted ecosystems.

Fire damages and kills plants via heating of living tissues, from exposed leaves and buds to cambial tissue and root systems (Hood et al. 2018). Damage to foliage, stems, and buds (collectively "total crown injury") are the most common post-fire predictors of tree mortality in empirical models (Woolley et al. 2012). This damage is often loosely defined as "crown scorch" and visually measured as either a percentage of total crown volume or total crown height (Hood et al. 2018). Many species, especially pines (*Pinus* spp.), can tolerate high levels of crown scorch (Fowler and Sieg 2004; Hood et al. 2007), with ample carbohydrate reserves to refoliate after the fire. Mortality due to cambial heating occurs when the cambium layer of the tree is killed around the entire circumference of the bole, girdling the tree. Mortality also occurs due to damage to tree roots, particularly due to long-duration smoldering of accumulated duff and bark slough around trees in long-unburned forests (Varner et al. 2007). These injuries may result in total tree mortality; however, many species (especially angiosperms) may only be top killed and eventually resprout from buds located either at the base of the tree or belowground as root sprouts or away from the stem as suckers.

Differential mortality occurs across species and across life stages within species due to traits that improve survivability. Traits such as the "grass stage" seedling physiognomy (Chapman 1932; Pile et al. 2017), rapid self-pruning (Rodríguez-Trejo and Fulé 2003; Pausas 2015), and thick outer bark (Hare 1965; Vines 1968; Hengst and Dawson 1993; Stephens and Libby 2006) protect meristems from the lethal heat energy doses. In some cases the trait is binary; the species either possesses the trait or does not (e.g., the grass stage or rapid self-pruning). Other traits such as thick bark can depend on the life stage of the species. Bark generally gets thicker as a tree grows for many species; however, the rate at which a species develops thick bark can differ (Agee 1996). Some species invest resources in thick bark early in life (negative allometry), others invest later (positive allometry), and some have a constant proportional allocation with increasing size (isometry, Jackson et al. 1999). In addition to the thickness of outer bark, observations about the roughness or rugosity of bark are currently being studied to determine if this trait can affect the boundary layer around the stem and provide a survival benefit during fire (Figure 6.3; O'Brien et al. 2018).

Figure 6.3 Example of bark rugosity between *Quercus laevis* (top right four cross-sections) and *Q. geminata* (bottom left).

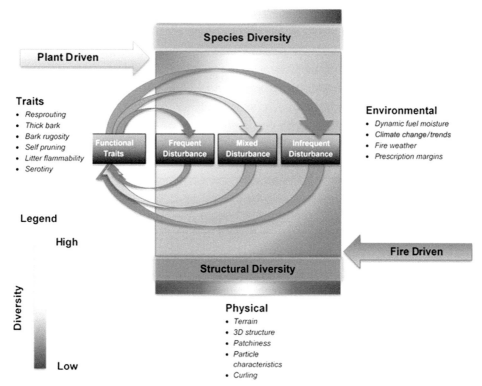

Figure 6.4 Synthetic model of plant–fire feedbacks. This conceptual model integrates the concepts of functional traits as a function of disturbance frequency (frequent, moderate, and rare). These functional traits are the evolutionary adaptations that plants develop to self-select for the type of fire and response that maintain their preferred disturbance regime. In this case, plant driven features are described as plants being the object and fire acting as the subject, where fire and plant response are a product of the plant adaptations. The reciprocal feedback to the plant driven model is a fire driven paradigm, where fire is the active object, generating larger fire effects that modify both structure and species diversity. These two endmembers are inextricably linked and feedback on each other over long temporal spans. Beyond functional traits, environmental and physical characteristics play a significant role in the ability for not only potential energy of fuels to be released, but also how, where, and when fire will affect ecosystems.
A black and white version of this figure will appear in some formats. For the color version, please refer to the plate section.

6.4 Unified Model of Plant and Fire Feedbacks

These case studies and the conceptual model we propose argue for a synthetic model of how plants and fire interact. Models of plant–fire feedbacks should acknowledge the wealth of information that fire and fuels science has accumulated on the role of proximal fuels on local fire behavior surrounding plants. We detail these aspects and note that others have reviewed how fuels are characterized by their dynamism (Keane 2015; Rowell et al. 2020). Plants, long an afterthought in these feedbacks, are the ultimate source of organic fuel, and research over the last two decades has shown tremendous variation in the flammability of their foliage, reproductive parts, and stems (Varner et al. 2015). Beyond "plants as fuel," the interaction between plant structure and how fire behaves is an overlooked aspect of plant–fire feedbacks. Our model includes plant structure (how boles interact with flows, how crowns cast shade and interact with flows) and how lacunarity ("where plants aren't") interacts with flows. Taken together, this synthetic view of plant–fire feedbacks (Figure 6.4) charts a new path in understanding how fires behave and the subsequent ecological consequences of those fires.

This synthetic model of plant–fire feedbacks is more than another abstraction, as it offers a way forward for physics-based models to characterize fuels, fire behavior, and fire effects. The core principles discussed thus far in this chapter come to play within this model, where the functional traits of plants influence the type of fire and response such that their preferred disturbance regime is maintained, which promotes a continuation of select species groupings. However, the role of fire as a disturbance agent also produces structure changes that may lead to a highly heterogeneous structure (e.g. mixed severity) or homogenous structures (e.g. rare disturbances). The multi-variate model proposed here integrates well with mechanisms and outputs that are inherent in computational fluid dynamics (CFD) models. We posit that evaluation of a range of scales used in a variety of CFD models may produce insights and the potential to quantify the agents that drive this reciprocal model. We discuss ideas for how to quantify and expand aspects of this conceptual physical considerations to inputs and outputs of CFD models may prove instrumental in providing innovative and informative methods of representing plant and fire feedbacks. The realization of this synthetic model of plant–fire feedbacks requires investment in quantification of key components that drive the model. As examples, mechanistic models of porous fuels metrics and fluid flow that generate multiscale plant–fire feedbacks that can be statistically compared with observed phenomena, improved representation of multiscale fuel moisture dynamics that describe how fuel potential energy is released, or CFD derived simulations of plant responses, whether simply survival and subsequent growth or as dramatic as resprouting or suckering could be included to encompass intervening plant community changes wrought by fire. The synthetic model of plant–fire feedbacks acts a module that not only links to the CFD model domain, but other fire–plant response models that are developing to characterize elements as ground species diversity in frequently burned systems (Loudermilk et al. 2019) under three distinct modalities of frequent, mixed, and infrequent occurrences of fire.

6.4.1 Frequently Burned Systems

Within frequent fire ecosystems such as those in subtropical pine forests or dry oak woodlands, fire acts as a selective filter, where species with traits to survive fire or traits to take advantage of the post-fire environment are favored and reinforced by fire. For example, frequent surface fires in many pine or oak-dominated ecosystems preferentially kill smaller diameter trees and shrubs and reduce forest floor material (litter and duff), favoring fire-resistant individuals and reducing surface fuels such that subsequent fires burn in leaf litter, pine needles, and/or shrub and herbaceous fuels. In frequently burned systems (with fire return intervals of 1–15 years depending on the ecosystem) plant functional traits generally maintain a system of low-intensity fire. Examples include needle characteristics in long needle pine species that support flammability (Fonda 2001), heat dose (Ellair and Platt 2013), and post-fire herbaceous understory regeneration that perpetuate frequent fire (Sackett et al. 1996). Structural adaptations such as self-pruning of the canopy limit impacts of frequent low intensity fire on the crown itself through creating gaps between canopy and surface fuels (Keeley 2012) and the ability to entrain cool air through large lacuna in the three-dimensional vegetation lattice. Fire-tolerating plant assemblages mix functional traits as thick bark, sprouting via apical sprouting (a protected apical bud that is insulated from the fire heat doses), basal buds resprouting (buds that are at or below ground that tap into lignotuberous carbohydrate reserves), below-ground bud banks (insulated by the soil and inherent in both woody and herbaceous species), and scarification of the seed bank via thermal dosing to promote germination (Cushwa et al. 1968; Agee 1998; Clarke et al. 2013; Varner et al. 2016).

The short interval between fires in frequent fire systems often suppresses understory development of shrubs and tree encroachment. Many of these systems can be described as open forests, savannas, or woodlands (low basal area or canopy cover) with a grass or herbaceous vegetation layer that is well insulated and responsive to fuel moisture changes (Anderson 1990; Nelson and Hiers 2008; Engber et al. 2011; Varner et al. 2015). If intervals between fires lengthen, the structural complexity of the stands can increase though mid-story tree encroachment (Glitzenstein et al. 1995) or shade tolerant species that benefit from dense overstory canopy cover that results in increasing structural stratification from expanding regeneration (Minnich et al. 1995; North et al. 2004; Engber et al. 2011).

Longleaf pine forests and understory plant communities of the southeastern United States are an archetypal representation of frequently burned systems. Factors such as the overstory composition and structure act on this system as objects that preferentially seed the fuelbed with elements that effectively carry fire (e.g. pine litter; Gresham 1982; O'Brien et al. 2016b). Openings and gaps in the canopy that promote grass and herbaceous features ultimately drive the ability for continued levels of fire that are supported by sufficient fuel loading (bulk density) and continuity (horizontal). In this context, fire is a feature of these species, preferentially generating the conditions needed to sustain fire return intervals of one to two years (Platt 1999; Glitzenstein et al. 2003; Reid and Robertson 2012). The morphology of longleaf pines includes evolutionary responses as thick bark to protect the cambium from thermal damage and self-pruning lower limbs from thermal doses over repeated fire, a fire survival strategy that allows the species to mitigate potential for fire transitioning to a crown fire (Keeley and Zedler 1998; Schwilk and Ackerly 2001; Keeley 2012). Frequent fires also moderate the ability for hardwood midstory species to attain statures that have the ability to inhibit the spread of fire under typical prescribed fire margins due to shading that spikes fuel moisture under the canopy, fuel properties of fire-dampening litter (e.g. SA:V, Kane et al. 2008; Kreye et al. 2013), and drag from structural porosity that moves fire around these objects producing fine-scale fire effects (O'Brien et al. 2016a).

An important aspect of frequently burned systems, such as longleaf pine or oak-dominated forests and savannas, is the cultural use of fire as the mechanism to maintain frequent fires and the resources and sustainable fire regime. Adaptation of the subgenus *Pinus* to tolerate frequent low intensity fire began long before cultural burning, but the role of human use of fire has carried this process to high levels of landscape application, where the pre-colonization period of the southeastern United States was dominated by fire-tolerant systems such as longleaf pine (Myers and White 1987; Millar 1998). Fowler and Konopik (2007) describe indigenous fire as the key form of disturbance throughout the southern United States for more than 10,000 years through five cultural periods.

6.4.2 Mixed Severity Fire

Mixed severity fires often describe the largest proportion of wildfire effects on dry mixed conifer forests of western US landscapes (Kane et al. 2015), with unburned, low, and moderate severity patches typically representing the majority of the burned area (Perry et al. 2011; Reilly et al. 2019). Through the burning and returning of forested landscapes, mixed severity fire regimes generally self-select for plant functional traits that are adapted to fire (e.g., thick-barked ponderosa, Jeffery pine, western larch and mature Douglas-fir) and in this way resemble frequent fire systems. However, stand-replacement events are also common within mixed-severity fire regimes and lead to a patchwork burn mosaic that has feedbacks to future fire events (Hessburg et al. 2019). Prior to fire exclusion, the majority of wildfires in dry mixed conifer forests were small- to medium-sized fires that created mosaics of forests across a range of ages and fuel loadings and on-forest vegetation, including grasslands and shrublands (Hagmann et al. 2019; Hessburg et al. 2019). This patchwork of forests and non-forests favored bottom-up controls of fire from patterns of vegetation and terrain (McKenzie and Littell 2017). However, climate-driven large fire events also occurred and would episodically burn across resilient patchworks, only to be rebuilt again by severity range in frequent small- to medium-sized fires.

Through their complex interactions with climate, terrain, and past disturbances, mixed severity fire regimes create a mosaic of three dimensional structural configurations that tend to limit subsequent fire spread and severity over forested landscapes and contribute to landscape heterogeneity (Figure 6.5) (Parks et al. 2014, 2015; Prichard et al. 2017; Hoff et al. 2019). For example, in US national parks and wilderness areas that allow unplanned ignitions to burn for resource benefit, burn mosaics have been demonstrated as short-term barriers to fire spread within longer term mitigation of fire severity (Parks et al. 2015, 2016, 2018; Prichard et al. 2017). Post-fire patterns of plant diversity follow a distribution that aligns the highest levels of diversity with low and low–moderate severity fire and reduced diversity in higher severity patches (DeSiervo et al. 2015; Stevens et al. 2015). The patchiness that ensues is supportive of both mesic and xeric habitats, thus adaptive and functional traits of a broad diversity are able to co-exist in close proximity (Stevens et al. 2015). As fires burn and reburn landscapes, edge environments, open forests dominated by fire-tolerant conifers and fire refugia become more dominant features across burned landscapes (Kolden et al. 2015; Krawchuk et al. 2020). Fire–atmosphere interactions within complex terrain and past burn mosaics often contribute to heterogeneous patch mosaics in mixed-severity fire regimes as the flow of fire and wind are mediated by topography and patches of forest and non-forest vegetation with different thresholds to burning.

The absence of fire in the twentieth century (Marlon et al. 2009) has led to profound changes in the continuity of western mixed conifer forests and their potential for high-intensity crown fire events (Hessburg et al. 2019). Over the past several decades, western wildfires have been increasing in area burned and extent of high severity effects (Parks and Abatzoglou 2020). Following nearly a century of fire exclusion in many western forests, first-entry fires can contribute to high severity reburns through development of flammable shrub fields. Standing dead trees can also dominate post-fire landscapes with a legacy of fire exclusion and contribute to high surface fuel loads and high severity reburns within regenerating forests (Kemp et al. 2016; Coop et al. 2020). However, if fires are allowed to burn and reburn over time, complex fire–vegetation interactions will gradually restore more resilient mosaics of forest, grasslands, and shrublands that tend to become more fuel and terrain limited (Prichard et al. 2017).

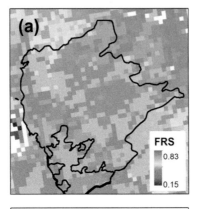

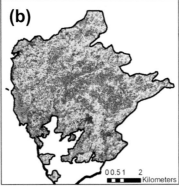

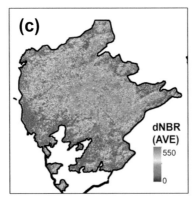

Figure 6.5 Patterns of mixed severity and disturbance example: North Rim Grand Canyon National Park, USA.

The north Rim of Grand Canyon National Park encompasses all three of the key disturbance type described in Figure 6.4. On lower peninsulas of the Kaibab Plateau, frequent fire in Ponderosa pine-gamble oak maintain functional traits as high lacunarity, thick bark for fire tolerance, and resprouting adaptations in deciduous shrubs after fire. At higher elevations, there are a mixture of mixed and rare disturbance regimes.

The Poplar fire occurred on the north Rim in 2003 in a matrix of mixed conifer and spruce-fire forest types burning 4,813 ha (Stoddard et al. 2020). The mixed

6.4.3 Infrequent Fire Systems

Forest ecosystems with infrequent fire regimes are generally characterized by stand-replacing fire and are often associated with thin-barked tree species that lack adaptations to frequent fire. However, labels such as "fire avoiders" miss some key fire adaptations and feedbacks. Where fire is a relatively rare occurrence, vegetation often exhibit fire embracing and avoiding mechanisms (Keeley 2001). For example, cold mixed conifer forests such as lodgepole pine, Engelmann spruce, and subalpine fir or spruce-dominated boreal forests have thin bark and crown architectures that predispose them to stand-replacing crown fires. In turn, stand-replacing fires create open light conditions and exposed mineral soils that are well suited for post-fire tree regeneration. Cold environments of montane, subalpine, and boreal landscapes often support high foliar moisture contents and mesic, shaded understories that contribute to long intervals between fires. Although infrequent fire systems tend to burn during exceptionally dry fire weather events and can be synchronous across regions (McKenzie et al. 2011), complex burn mosaics are supported by fires burning and reburning these landscapes. Although infrequent fire ecosystems may at first seem less influenced by fire, legacies of past fires still have pronounced impacts on vegetation trajectories and feedbacks to subsequent fires. Fire–vegetation feedbacks within rare disturbance regimes are therefore best evaluated at broad landscape scales as the patterns of past fires influence the subsequent spread and severity of fires. Repeat fires within cold forest systems can also lead to alternative vegetation states, including perpetuation of high-elevation meadows (Cansler et al. 2018) or dominance of broadleaf deciduous forests (Hart et al. 2019).

Caption for figure 6.5 (*cont.*). severity Poplar fire burned through a broad range of fire resilience scores, an integrated metric of functional traits that quantifies the ability of the conifer system to maintain its current state after fire (a) (Stevens et al. 2020). Airborne laser scanning (ALS) derived structure classification from horizontal, vertical, and surface roughness metrics (Hoff et al. 2019) describes the structural variability resulting from the high severity patches (b), green and blue colors) in areas with low-to-moderate fire resilience scores. These sites represent state changes as a result of the high severity from mixed conifer and spruce-fire to resprouting aspen and New Mexico locust or grass land.

Comparison with the delta normalized burn ratio (dNBR) metric from the Monitoring Trends in Burn Severity program (MTBS; www.mtbs.gov (last accessed November 20, 2021)) with the ALS derived structure classes (c) shows predictable structural outcomes from heat flux, where fire resilient structure types that employ functional traits have porosity and lacunarity metrics that move convective heat efficiently.

A black and white version of this figure will appear in some formats. For the color version, please refer to the plate section.

6.5 Future Directions and Applications

Our conceptual model (Figure 6.4) highlights several directions for future research. Currently, fire research that integrates fuels, fire behavior, and fire effects are rare (Hood et al. 2018; O'Brien et al. 2018). While plant flammability has been evaluated in laboratory settings, in situ measurements are needed to evaluate how fuel structure, composition, and physical properties interact with fire. Specifically, bark and bole structure, crown and canopy structure, and lacunarity deserve similar investigation to determine their relative influence on fire spread, intensity, and duration, and where important vegetation–fire interactions may exist. As described in our conceptual model of fire–vegetation interactions, vegetation contributes to both the availability of fuels for combustion and as objects, and influences the flow of fire spread and wind through porous media, gaps in available fuels, and around barriers (e.g., tree boles).

As we addressed in Section 6.2, short-term changes in fuel moisture and structure (4-D fuels) lack substantial attention, but variations in fuel moisture across fine temporal scales (i.e., sub-hourly to hourly) are key to fire behavior. In particular, live fuel moisture can dictate how readily fires can ignite and spread through vegetation. As such, living plants can present barriers to fire spread or heat sinks that reduce potential fire spread and intensity. Models that incorporate plant physiology, live and dead fuel moisture dynamics in plants (living and dead), and thresholds to availability for combustion are needed. Applications of these designs should attempt to mesh with ongoing computational fluid dynamics models that can improve our understanding of how fires behave and why they behave the way they do. The ultimate goal of these efforts should be to re-envision the ecology of fuels, and how fire and vegetation interact across a range of spatial and temporal scales.

References

Agee, JK (1993) Methods of evaluating forest fire history. *Journal of Northeast Forestry University* 4(2), 1–10.

Agee, JK (1996) *Fire Ecology of Pacific Northwest Forests.* Washington, DC: Island Press.

Agee, JK (1998) The landscape ecology of western forest fire regimes. *Northwest Science* 72(Special Issue), 24–34.

Agee, JK, Bahro, B, Finney, MA, Omi, N, Sapsis, DB, Skinner, CN, Van Wagtendonk, JW, Weatherspoon, CP (2000) The use of shaded fuelbreaks in landscape fire management. *Forest Ecology and Management* 127(1–3), 55–66.

Agee, JK, Skinner, CN (2005) Basic principles of forest fuel reduction treatments. *Forest Ecology and Management* 211(1–2), 83–96.

Anderson, HE (1969) *Heat Transfer and Fire Spread.* Ogden, UT: US Department of Agriculture, Forest Service, Intermountain Forest and Range Experiment Station.

Anderson, HE (1970) Forest fuel ignitability. *Fire Technology* 6, 312–319.

Anderson, HE (1982) *Aids to Determining Fuel Models for Estimating Fire Behavior.* Ogden, UT: US Department of Agriculture, Forest Service, Intermountain Forest and Range Experiment Station.

Anderson, HE (1990) *Predicting Equilibrium Moisture Content of Some Foliar Forest Litter in the Northern Rocky Mountains.* Research Paper – US Department of Agriculture, Forest Service, no. INT-429.

Banwell, EM, Varner, JM, Knapp, EE, Van Kirk, RW (2013) Spatial, seasonal, and diel forest floor moisture dynamics in Jeffrey pine–white fir forests of the Lake Tahoe Basin, USA. *Forest Ecology and Management* **305**(October), 11–20.

Bond, WJ, Keeley, JE (2005) Fire as a global "Herbivore": The ecology and evolution of flammable ecosystems. *Trends in Ecology and Evolution* **20**(7), 387–394.

Burgan, RE, Rothermel, RC (1984) *BEHAVE: Fire Behavior Prediction and Fuel Modeling System: FUEL Subsystem.* General Technical Report INT-167. Ogden, UT: US Department of Agriculture, Forest Service, Intermountain Forest and Range Experiment Station.

Byram, GM (1959) Combustion of forest fuels. In: Davis, KP, ed. *Synthesis of Knowledge of Extreme Fire Behavior: Volumie I for Fire Managers.* New York: McGraw-Hill, pp. 61–89.

Cansler, CA, McKenzie, D, Halpern, CB (2018) Fire enhances the complexity of forest structure in alpine treeline ecotones. *Ecosphere* **9**(2), 1–21.

Chapman, HH (1932) Is the longleaf type a climax? *Ecology* **13**(4), 328–334.

Chrosciewicz, Z (1986) Foliar moisture content variations in four coniferous tree species of central Alberta. *Canadian Journal of Forest Research* **16**(1), 157–162.

Clarke, PJ, Lawes, MJ, Midgley, JJ, Lamont, BB, Ojeda, F, Burrows, GE, Enright, NJ, Knox, KJE (2013) Resprouting as a key functional trait: How buds, protection and resources drive persistence after fire. *New Phytologist* **197**(1), 19–35.

Coop, JD, Parks, SA, Stevens-Rumann, CS, Crausbay, SD, Higuera, PE, Hurteau, MD, Tepley, A, Whitman, E, Assal, T, Collins, BM, Davis, KT, Dobrowski, S, Falk, DA, Fornwalt, PJ, Fulé, PZ, Harvey, BJ, Kane, VR, Littlefield, CE, Margolis, EQ, North, M, Parisien, MA, Prichard, S, Rodman, KC (2020) Wildfire-driven forest conversion in Western North American landscapes. *BioScience* **70**(8), 659–673.

Cushwa, CT, Martin, RE, Miller, RL (1968) The effects of fire on seed germination. *Journal of Range Management* **21**, 250–254.

Delagrange, S, Potvin, C, Messier, C, Coll, L (2008) Linking multiple-level tree traits with biomass accumulation in native tree species used for reforestation in Panama. *Trees – Structure and Function* **22**(3), 337–349.

DeSiervo, MH, Jules, ES, Safford, HD (2015) Disturbance response across a productivity gradient: Postfire vegetation in serpentine and nonserpentine forests. *Ecosphere* **6**(4), 1–19.

Ellair, DP, Platt, WJ (2013) Fuel composition influences fire characteristics and understorey hardwoods in pine savanna. *Journal of Ecology* **101**(1), 192–201.

Emery, N, Roth, K, Pivovaroff, AL (2020) Flowering phenology indicates plant flammability in a dominant shrub species. *Ecological Indicators* **109**, 105745.

Engber, EA, Varner, JM (2012) Patterns of flammability of the California oaks: The role of leaf traits. *Canadian Journal of Forest Research* **42**(11), 1965–1975.

Engber, EA, Varner, JM, Arguello, LA, Sugihara, NG (2011) The effects of conifer encroachment and overstory structure on fuels and fire in an oak woodland landscape. *Fire Ecology* **7**(2), 32–50.

Fill, JM, Moule, BM, Varner, JM, Mousseau, TA (2016) Flammability of the keystone savanna bunchgrass *Aristida stricta*. *Plant Ecology* **217**(3), 331–342.

Fonda, RW (2001) Burning characteristics of needles from eight pine species. *Forest Science* **47**(2), 390–396.

Fosberg, MA, Lancaster, JW, Schroeder, MJ (1970) Fuel moisture response: Drying relationships under standard and field conditions. *Forest Science* **16**(1), 121–128.

Fowler, C, Konopik, E (2007) The history of fire in the southern United States. *Human Ecology Review* **14**(2), 165–176.

Fowler, JF, Sieg, CH (2004) *Postfire Mortality of Ponderosa Pine and Douglas-Fir: A Review of Methods to Predict Tree Death*. USDA Forest Service: General Technical Report RMRS-GTR, no. 132 RMRS-GTR: 1–27.

Glitzenstein, JS, Platt, WJ, Streng, DR (1995) Effects of fire regime and habitat on tree dynamics in North Florida longleaf pine savannas. *Ecological Monographs* **65**(4), 441.

Glitzenstein, JS, Streng, DR, Wade, DD (2003) Fire frequency effects on longleaf pine (*Pinus Palustris* P. Miller) vegetation in South Carolina and Northeast Florida, USA. *Natural Areas Journal* **23**(1), 22–37.

Gresham, CA (1982) Litterfall patterns in mature loblolly and longleaf pine stands in coastal South Carolina. *Forest Science* **28**(2), 223–231.

Grimm, V (1994) Mathematical models and understanding in ecology. *Ecological Modelling* **75–76**(September), 641–651.

Gruell, GE, Brown, JK, Bushey, CL (1986) *Prescribed Fire Opportunities in Grasslands Invaded by Douglas-Fir: State-of-the-Art Guidelines*. General Technical Report, Intermountain Research Station, USDA Forest Service No. INT-19: 19.

Gutsell, SL, Johnson, EA (1996) How fire scars are formed: Coupling a disturbance process to its ecological effect. *Canadian Journal of Forest Research* **26**(2), 166–174.

Hagmann, RK, Merschel, AG, Reilly, MJ (2019) Historical patterns of fire severity and forest structure and composition in a landscape structured by frequent large fires: Pumice Plateau ecoregion, Oregon, USA. *Landscape Ecology* **34**, 551–568.

Hare, RC (1965) Contribution of bark to fire resistance of southern trees. *Journal of Forestry* **63**(4), 248–251.

Hart, SJ, Henkelman, J, McLoughlin, PD, Nielsen, SE, Truchon-Savard, A, Johnstone, JF (2019) Examining forest resilience to changing fire frequency in a fire-prone region of boreal forest. *Global Change Biology* **25**(3), 869–884.

Hengst, GE, Dawson, JO (1993) Bark thermal properties of selected central hardwood species. In North Central Forest Experiment Station, Forest Service, U.S. Dept. of Agriculture*, 28:5241–44. 9th Central Hardwood Forest Conference: Proceedings of a Meeting Held at Purdue Univeristy* [sic], West Lafayette, IN, March 8–10. St. Paul, MN: North Central Forest Experiment Station, Forest Service, U.S. Dept. of Agriculture. p. 5.

Hermann, SH (1993) The longleaf pine ecosystem: Ecology, restoration and management. In *Proceedings of the 18th Tall Timbers Fire Ecology Conference*, May 30–June 2, 1991, Tallahassee, FL.

Hessburg, PF, Miller, CL, Parks, SA, Povak, NA, Taylor, AH, Higuera, PE, Prichard, SJ, North, MP, Collins, BM, Hurteau, MD, Larson, AJ, Allen, CD, Stephens, SL, Rivera-Huerta, H, Stevens-Rumann, CS, Daniels, LD, Gedalof, Z, Gray, RW, Kane, VR, Churchill, DJ, Hagmann, RK, Spies, TA, Cansler, CA, Belote, RT, Veblen, TT, Battaglia, MA, Hoffman, C, Skinner, CN, Safford, HD, Salter, RM (2019) Climate, environment, and disturbance history govern resilience of western North American forests. *Frontiers in Ecology and Evolution* **7**(July), 239.

Hiers, JK, O'Brien JJ, Mitchell RJ, Grego JM, Loudermilk EL (2009) The wildland fuel cell concept: An approach to characterize fine-scale variation in fuels and fire in frequently burned longleaf pine forests. *International Journal of Wildland Fire* **18**(3), 315–325.

Hoff, V, Rowell, E, Teske, C, Queen, L, Wallace, T (2019) Assessing the relationship between forest structure and fire severity on the north rim of the Grand Canyon. *Fire* **2**(1), 1–22.

Hoffman, CM, Linn, R, Parsons, R, Sieg, C, Winterkamp, J (2015) Modeling spatial and temporal dynamics of wind flow and potential fire behavior following a mountain pine beetle outbreak in a lodgepole pine forest. *Agricultural and Forest Meteorology* **204**, 79–93.

Hoffmann, WA, Geiger, EL, Gotsch, SG, Rossatto, DR, Silva, LCR, Lau, OL, Haridasan, M, Franco, AC (2012) Ecological thresholds at the savanna–forest boundary: How plant traits, resources and fire govern the distribution of tropical biomes. *Ecology Letters* **15**(7), 759–768.

Hoffmann, WA, Orthen, B, Kielse, P, Nascimento, VDO (2003) Comparative fire ecology of tropical savanna and forest trees. *Functional Ecology* **17**(6), 720–726.

Hood, SM, McHugh, CW, Ryan, KC, Reinhardt, E, Smith, SL (2007) Evaluation of a post-fire tree mortality model for Western USA conifers. *International Journal of Wildland Fire* **16**(6), 679–689.

Hood, SM, Varner, JM, Van Mantgem, P, Cansler, CA (2018) Fire and tree death: Understanding and improving modeling of fire-induced tree mortality. *Environmental Research Letters* **13**(11), 1–17.

Jackson, JF, Adams, DC, Jackson, UB (1999) Allometry of constitutive defense: A model and a comparative test with tree bark and fire regime. *American Naturalist* **153**(6), 614–632.

Jolly, WM, Hadlow, AM, Huguet, K (2014) De-coupling seasonal changes in water content and dry matter to predict live conifer foliar moisture content. *International Journal of Wildland Fire* **23**(4), 480–489.

Kane, JM, Varner, JM, Hiers, J (2008) The burning characteristics of southeastern oaks: Discriminating fire facilitators from fire impeders. *Forest Ecology and Management* **256**(12), 2039–2045.

Kane, VR, Lutz, JA, Cansler, CA, Povak, NA, Churchill, DJ, Smith, DF, Kane, JT, North, MP (2015) Water balance and topography predict fire and forest structure patterns. *Forest Ecology and Management* **338**, 1–13.

Keane, RE (2015) *Wildland Fuel Fundamentals and Applications*. Cham: Springer International Publishing.

Keeley, JE (2001) Fire and invasive species in Mediterranean-climate ecosystems of California. In Galley KEM, Wilson, TP, eds. *Proceedings of the Invasive Species Workshop: The Role of Fire in the Control and Spread of Invasive Species*. Miscellaneous Publication Number 11. Tallahassee, FL: Tall Timbers Research Station, pp. 81–94.

Keeley, JE (2012) Ecology and evolution of pine life histories. *Annals of Forest Science* **69**(4), 445–453.

Keeley, JE, Zedler, PH (1998) Evolution of life histories in *Pinus* . In: Richardson, D, ed. *Ecology and Biogeography of Pines*. Cambridge: Cambridge University Press, pp. 219–251.

Kemp, KB, Higuera, PE, Morgan, P (2016) Fire legacies impact conifer regeneration across environmental gradients in the U.S. Northern Rockies. *Landscape Ecology* **31**(3), 619–636.

Kolden, CA, Abatzoglou, JT, Lutz, JA, Cansler, CA, Kane, JT, Van Wagtendonk, JW, Key, CH (2015) Climate contributors to forest mosaics: Ecological persistence following wildfire. *Northwest Science* **89**(3), 219–238.

Krawchuk, MA, Meigs, GW, Cartwright, JM, Coop, JD, Davis, R, Holz, A, Kolden, C, Meddens. AJH (2020) Disturbance refugia within mosaics of forest fire, drought, and insect outbreaks. *Frontiers in Ecology and the Environment* **18**(5), 235–244.

Krawchuk, MA, Moritz, MA, Parisien, M-A, Van Dorn, J, Hayhoe, K (2009) Global pyrogeography: The current and future distribution of wildfire. *PLoS ONE* **4**(4), e5102.

Kreye, JK, Varner, JM, Hamby, GW, Kane. JM (2018) Mesophytic litter dampens flammability in fire-excluded pyrophytic oak–hickory woodlands. *Ecosphere* **9**(1), e02078.

Kreye, JK, Varner, JM, Hiers, JK, Mola, J (2013) Toward a mechanism for eastern North American forest mesophication: differential litter drying across 17 species. *Ecological Applications* **23**(8), 1976–1986.

Linn, RR, Cunningham, P (2005) Numerical simulations of grass fires using a coupled atmosphere–fire model: Basic fire behavior and dependence on wind speed. *Journal of Geophysical Research* **11**(1–14), 1–67.

Linn, RR, Sieg, CH, Hoffman, CM, Winterkamp, JL, McMillin, JD (2013) Modeling wind fields and fire propagation following bark beetle outbreaks in spatially-heterogeneous pinyon-juniper woodland fuel complexes. *Agricultural and Forest Meteorology* **173**, 139–153.

Little, CHA (1970) Seasonal changes in carbohydrate and moisture content in needles of balsam fir (Abies Balsamea). *Canadian Journal of Botany* **48**(11), 2021–2028.

Loudermilk, EL, Dyer, L, Pokswinski, S, Hudak, AT, Hornsby, B, Richards, L, Dell, J, Goodrick, SL, Hiers, JK, O'Brien, JJ (2019) Simulating groundcover community assembly in a frequently burned ecosystem using a simple neutral model. *Frontiers in Plant Science* **10**(September), 1107.

de Magalhães, RMQ, Schwilk, DW (2012) Leaf traits and litter flammability: Evidence for non-additive mixture effects in a temperate forest. *Journal of Ecology* **100**(5), 1153–1163.

Mäkelä, A (1997) A carbon balance model of growth and self-pruning in trees based on structural relationships. *Forest Science* **43**(1), 7–24.

Mandelbrot, BB (1983) *The Fractal Geometry of Nature*, Revised and Enlarged ed. New York: IBM Thomas J. Watson Research Center.

Marlon, JR, Bartlein, PJ, Walsh, MK, Harrison, SP, Brown, KJ, Edwards, ME, Higuera, PE, Power, MJ, Anderson, RS, Briles, C, Brunelle, A, Carcaillet, C, Daniels, M, Hu, FS, Lavoie, M, Long, C, Minckley, T, Richard, PJH, Scott, AC, Shafer, DS, Tinner, W, Umbanhowar, CE, Whitlock, C (2009) Wildfire responses to abrupt climate change in North America. *Proceedings of the National Academy of Sciences (USA)* **106**(8), 2519–2524.

Marshall, G, Thompson, DK, Anderson, K, Simpson, B, Linn, R, Schroeder, D (2020) The impact of fuel treatments on wildfire behavior in North American boreal fuels: A simulation study using FIRETEC. *Fire* **3**(2), 1–14.

McKenzie, D, Littell, J (2017) Climate change and the eco-hydrology of fire: Will area burned increase in a warming western USA? *Ecological Applications* **27**(1), 26–36.

McKenzie, D, Miller, C, Falk, DA (2011) *The Landscape of Ecology of Fire*. New York: Springer Science and Media.

McLauchlan, KK, Higuera, PE, Miesel, J, Rogers, BM, Schweitzer, J, Shuman, JK, Tepley, AJ, Varner, JM, Veblen, TT, Adalsteinsson, SA, Balch, JK, Baker, P, Batllori, E, Bigio, E, Brando, P, Cattau, M, Chipman, ML, Coen, J, Crandall, R,

Daniels, L, Enright, N, Gross, WS, Harvey, BJ, Hatten, JA, Hermann, S, Hewitt, RE, Kobziar, LN, Landesmann, JB, Loranty, MM, Maezumi, SY, Mearns, L, Moritz, M, Myers, JA, Pausas, JG, Pellegrini, AFA, Platt, WJ, Roozeboom, J, Safford, H, Santos, F, Scheller, RM, Sherriff, RL, Smith, KG, Smith, MD, Watts, AC (2020) Fire as a fundamental ecological process: Research advances and frontiers. *Journal of Ecology* **108**(5), 2047–2069.

Mell, W, Jenkins, MA, Gould, J, Cheny, P (2007) A physics based approach to modeling grassland fires. *International Journal of Wildand Fire* **16**(1), 1.

Millar, CI (1998) Early evolution of pines. In: *Ecology and Biogeography of Pinus*. Cambridge: Cambridge University Press.

Minnich, RA, Barbour, MG, Burk, JH, and Fernau, RF (1995) Sixty years of change in Californian conifer forests of the San Bernardino Mountains. *Conservation Biology* **9**(4), 902–914.

Myers, RL, White, DL (1987) Landscape history and changes in sandhill vegetation in north-central and south-central Florida. *Bulletin of the Torrey Botanical Club* **114**(1), 21.

Nelson, RM, Hiers, JK (2008) The influence of fuelbed properties on moisture drying rates and timelags of longleaf pine litter. *Canadian Journal of Forest Research* **38**(9), 2394–2404.

North, M, Chen, J, Oakley, B, Song, B, Rudnicki, M, Gray, A, Innes, J (2004) Forest stand structure and pattern of old-growth western hemlock/douglas-fir and mixed-conifer forests. *Forest Science* **50**(3), 299–311.

O'Brien, JJ, Hiers, JK, Varner, JM, Hoffman, CM, Dickinson, MB, Michaletz, ST, Loudermilk, EL, Butler, BW (2018) Advances in mechanistic approaches to quantifying biophysical fire effects. *Current Forestry Reports* **4**(4), 161–177.

O'Brien, JJ, Loudermilk, EL, Hornsby, B, Hudak, AT, Bright, BC, Dickinson, MB, Hiers, JK, Teske, C, Ottmar, RD (2016a) High-resolution infrared thermography for capturing wildland fire behavior – RxCADRE 2012. *International Journal of Wildland Fire* **25**(1), 62–75.

O'Brien, JJ, Loudermilk, EL, Hornsby, B, Polswinski, S, Hudak, A, Strother, RE, Bright, B (2016b) Canopy derived fuels drive patterns of in-fire energy release and understory plant mortality in a longleaf pine (*Pinus Palustris*) sandhill in Northwest FL, USA. *Canadian Journal of Remote Sensing* **42**(5), 489–500.

Ottmar, RD, Burns, MF, Hall, JN, Hanson, AD (1992) *CONSUME [1.0] Users Guide*. General Technical Report PNW-GTR-304. Portland, OR: USDA Forest Service, Pacific Northwest Research Station.

Ottmar, RD, Sandberg, DV, Riccardi, CL, Prichard, SJ (2007) An overview of the fuel characteristic classification system: Quantifying, classifying, and creating fuelbeds for resource planning. *Canadian Journal of Forest Research* **37**, 2383–2393.

Pallardy, SG (2008) *Physiology of Woody Plants*, 3rd ed. Amsterdam: Academic Press.

Parks, SA, Abatzoglou. JT (2020) Warmer and drier fire seasons contribute to increases in area burned at high severity in Western US forests from 1985–2017. *Geophysical Research Letters* **47**(22), e2020GL089858.

Parks, SA, Holsinger, LM, Miller, C, Nelson, CR (2015) Wildland fire as a self-regulating mechanism: The role of previous burns and weather in limiting fire progression. *Ecological Applications* **25**(6), 1478–1492.

Parks, SA, Holsinger, LM, Voss, MA, Loehman, RA, Robinson, NP (2018) Mean composite fire severity metrics computed with Google Earth engine offer improved accuracy and expanded mapping potential. *Remote Sensing* **10**(6), 1–15.

Parks, SA, Miller, C, Holsinger, LM, Baggett, LS, Bird, BJ (2016) Wildland fire limits subsequent fire occurrence. *International Journal of Wildland Fire* **25**(2), 182–190.

Parks, SA, Miller, C, Nelson, CR, Holden, ZA (2014) Previous fires moderate burn severity of subsequent wildland fires in two large western US wilderness areas. *Ecosystems* **17**(1), 29–42.

Parsons, RA, Mell, WE, McCauley, P (2011) Linking 3D spatial models of fuels and fire: Effects of spatial heterogeneity on fire behavior. *Ecological Modelling* **222**(3), 679–691.

Pausas, JG (2015) Bark thickness and fire regime. *Functional Ecology* **29**(3), 315–327.

Pausas, JG, Bradstock, RA (2007) Fire persistence traits of plants along a productivity and disturbance gradient in Mediterranean shrublands of South-East Australia. *Global Ecology and Biogeography* **16**(3), 330–340.

Pausas, JG, Paula, S (2012) Fuel shapes the fire–climate relationship: Evidence from Mediterranean ecosystems. *Global Ecology and Biogeography* **21**(11), 1074–1082.

Perry, DA, Hessburg, PF, Skinner, CN, Spies, TA, Stephens, SL, Taylor, AH, Franklin, JF, McComb, B, Riegel, G (2011) The ecology of mixed severity fire regimes in Washington, Oregon, and Northern California. *Forest Ecology and Management* **262**(5), 703–717.

Pickett, BM, Isackson, C, Wunder, R, Fletcher, TH, Butler, BW, Weise, DR (2009) Flame interactions and burning characteristics of two live leaf samples. *International Journal of Wildland Fire* **18**(7), 865.

Pickett, BM, Isackson, C, Wunder, R, Fletcher, TH, Butler, BW, Weise, DR (2010) Experimental measurements during combustion of moist individual foliage samples. *International Journal of Wildland Fire* **19**(2), 153–162.

Pile, LS, Wang, GG, Knapp, BO, Liu, G, Yu, D (2017) Comparing morphology and physiology of southeastern US *Pinus* seedlings: implications for adaptation to surface fire regimes. *Annals of Forest Science* **74**, 68.

Pimont, F, Dupuy, JL, Linn, RR, Dupont, S (2011) Impacts of tree canopy structure on wind flows and fire propagation simulated with FIRETEC. *Annals of Forest Science* **68**(3), 523–530.

Platt, WJ (1999) Southeastern pine savannas. In: Anderson, RC, Fralish, JS, Baskin, JM, eds. *Savannas, Barrens, and Rockoutcrop Plant Communities of North America.* Cambridge: Cambridge University Press.

Prichard, SJ, Kennedy, MC, Wright, CS, Cronan, JB, Ottmar, RD (2017) Predicting forest floor and woody fuel consumption from prescribed burns in southern and western pine ecosystems of the United States. *Data in Brief* **15**, 742–746.

Prichard, SJ, Ottmar, RD, Anderson, GK (2006) *Consume 3.0 User's Guide.* Seattle, WA: USDA Forest Service, Pacific Northwest Research Station, Pacific Wildland Fire Sciences Laboratory, Fire and Environmental Research Applications Team.

Reeves, MC, Ryan, KC, Rollins, MG, Thompson, TG (2009) Spatial fuel data products of the LANDFIRE project. *International Journal of Wildland Fire* **18**(3), 250–267.

Reid, AM, Robertson, KM (2012) Energy content of common fuels in upland pine savannas of the south-eastern US and their application to fire behaviour modelling. *International Journal of Wildland Fire* **21**(5), 591–595.

Reilly, MJ, Monleon, VJ, Jules, ES, Butz, RJ (2019) Range-wide population structure and dynamics of a serotinous conifer, knobcone pine (*Pinus Attenuata* L.), under an anthropogenically-altered disturbance regime. *Forest Ecology and Management* **441**(March), 182–191.

Reinhardt, E, Keane, RE, Brown, JK (1997) *First Order Fire Effects Model: FOFEM 4.0, User's Guide.* United States Department of Agriculture, Forest Service, Intermountain Research Station, General Technical Report INT-GTR-344.

Renninger, HJ, Carlo, NJ, Clark, KL, Schäfer, KVR (2015) Resource use and efficiency, and stomatal responses to environmental drivers of oak and pine species in an Atlantic coastal plain forest. *Frontiers in Plant Science* **6**, 297.

Rodríguez-Trejo, DA, Fulé, PZ (2003) Fire ecology of Mexican pines and a fire management proposal. *International Journal of Wildland Fire* **12**(1), 23–37.

Rollins, MG (2009) LANDFIRE: A nationally consistent vegetation, wildland fire, and fuel assessment. *International Journal of Wildland Fire* **18**(3), 235–249.

Rothermel, RC (1972) *A Mathematical Model for Predicting Fire Spread in Wildland Fuels.* Research Paper, INT-115. Ogden, UT: US Department of Agriculture, Intermountain Forest and Range Experiment Station.

Rothermel, RC (1991) *Predicting Behavior and Size of Crown Fires in the Northern Rocky Mountains.* USDA Forest Service, Intermountain Research Station, Research Paper, no. January: 46.

Rowell, E, Loudermilk, EL, Hawley, C, Pokswinski, S, Seielstad, C, Queen, L, O'Brien, JJ, Hudak, AT, Goodrick, S, Hiers, JK (2020) Coupling terrestrial laser scanning with 3D fuel biomass sampling for advancing wildland fuels characterization. *Forest Ecology and Management* **462**(April), 117945.

Sackett, SS, Haase, SM, Harrington, MG (1996) Lessons learned from fire use for restoring southwestern Ponderosa pine ecosystems. In Covington, WW, Wagner, PK, eds., *Conference on Adaptive Ecosystem Restoration and Management: Restoration of Codilleran Conifer Landscapes of North America*, June 6–8, 1996, Flagstaff, AZ. General Technical Report GTR-RM-278. Fort Collins, CO, USDA Forest Service, Rocky Mountain Forest and Range Experiment Station: 54–61.

Sandberg, DV, Riccardi, CL, Schaaf, MD (2007) Fire potential rating for wildland fuelbeds using the fuel characteristic classification system. *Canadian Journal of Forest Research* **37**(12), 2456–2463.

Schwilk, DW, Ackerly, DD (2001) Flammability and serotiny as strategies: Correlated evolution in pines. *Oikos* **94**(2), 326–336.

Scott, JH, Burgan, RE (2005) *Standard Fire Behavior Fuel Models: A Comprehensive Set for Use with Rothermel's Surface Fire Spread Model.* USDA Forest Service – General Technical Report RMRS-GTR, no. 153 RMRS-GTR: 1–76.

Spearpoint, MJ (1999) *Predicting the Ignition and Burning Rate of Wood in the Cone Calorimeter Using an Integral Model.* Gaitherburg, MD: National Institute of Standards and Technology.

Stephens, SL, Libby, WJ (2006) Anthropogenic fire and bark thickness in coastal and island pine populations from Alta and Baja, California. *Journal of Biogeography* **33**(4), 648–652.

Stevens, JT, Kling, MM, Schwilk, DW, Varner, JM, Kane, JM (2020) Biogeography of fire regimes in Western U.S. conifer forests: A trait-based approach. *Global Ecology and Biogeography* **29**(5), 944–955.

Stevens, JT, Safford, HD, Harrison, S, Latimer, AM (2015) Forest disturbance accelerates thermophilization of understory plant communities. *Journal of Ecology* **103**(5), 1253–1263.

Stoddard, MT, Fulé, PZ, Huffman, DW, Sánchez Meador, AJ, Roccaforte, JP (2020) Ecosystem management applications of resource objective wildfires in forests of the Grand Canyon National Park, USA. *International Journal of Wildland Fire* **29**(2), 190–200.

Van Wagner, CE (1975) "Convection temperatures above low intensity forest fires." *Bi-Monthly Research Notes Canadian Forest Service* **31**(2), 21.

Van Wagner, CE (1977) Conditions for the start and spread of crown fire. *Canadian Journal of Forest Research* **7**(1), 23–34.

Varner, JM, Hiers, JK, Ottmar, RD, Gordon, DR, Putz, FE, Wade, DD (2007) Overstory tree mortality resulting from reintroducing fire to long-unburned longleaf pine forests: The importance of duff moisture. *Canadian Journal of Forest Research* **37**(8), 1349–1358.

Varner, JM, Kane, JM, Hiers, JK, Kreye, JK, Veldman, JW (2016) Suites of fire-adapted traits of oaks in the Southeastern USA: Multiple strategies for persistence. *Fire Ecology* **12**(2), 48–64.

Varner, JM, Kane, JM, Kreye, JK, Engber, E (2015) The flammability of forest and woodland litter: A synthesis. *Current Forestry Reports* **1**(2), 91–99.

Vincent, G, Harja, D (2008) Exploring ecological significance of tree crown plasticity through three-dimensional modelling. *Annals of Botany* **101**(8), 1221–1231.

Vines, RG (1968) Heat transfer through bark, and the resistance of trees to fire. *Australian Journal of Botany* **16**(3), 499.

Viney, NR (1991) A review of fine fuel moisture modelling. *International Journal of Wildland Fire* **1**(4), 215.

Wang, GG (2002) Fire severity in relation to canopy composition within burned boreal mixedwood stands. *Forest Ecology and Management* **163**, 85–92.

Waring, RH, Running, SW (1998) *Forest Ecosystems: Analysis at Multiple Scales.* New York: Academic Press.

Williamson, GB, Black, EM (1981) High temperature of forest fires under pines as a selective advantage over oaks. *Nature* **293**(5834), 643–644.

Woolley, T, Shaw, DC, Ganio, LM, Fitzgerald, S (2012) A review of logistic regression models used to predict post-fire tree mortality of western North American conifers. *International Journal of Wildland Fire* **21**(1), 1.

Zeide, B (1993) Analysis of growth equations. *Forest Science* **39**(3), 594–616.

7

Wind and Canopies

FRANÇOIS PIMONT, JEAN-LUC DUPUY, AND RODMAN LINN

7.1 Introduction

Wind is one of the key factors affecting fire behavior, through several main mechanisms involved in fire spread. For example, the radiative transfer is influenced by flame inclination, which angle is a function of wind speed (Albini 1985); at the same time, the wind flow advects the hot gases that are involved in convective heat transfer to the solid fuel (Baines 1990); finally, the turbulence of the wind flow increases the mixing involved in both convection and combustion through complex fluid dynamics mechanisms (Tieszen 2001). Observations and experimental data suggest that the rate of spread of a fire is close to being proportional to wind speed (e.g. Cheney et al. 1998), therefore fire intensity also increases linearly with wind speed to a first order. Wind speed also affects the transition to crowning through surface fire intensity, as well as the crown fire rate of spread (Van Wagner 1977; Cruz et al. 2005). It is, hence, a major factor in most fire danger metrics, usually expressed as the "fire wind" (Andrews 2012), which is defined either at mid-flame height or at a standardized height (usually 10 m or 20 ft above ground level). A popular example of such metric is the Fire Weather Index (FWI), in which a subcomponent called the Initial Spread Index (ISI) is an exponential function of fire wind speed. At landscape scale, observations of fire activity show that daily wind speed strongly affects the probability of fire occurrence and increases fire size, by shifting the burned-area distribution to larger areas (Hernandez et al. 2015).

In the lower part of the atmosphere, wind flows are essentially driven by mesoscale pressure gradients. They are also affected by the Coriolis force, which results from Earth's rotation, and by interactions with the features at the surface (soil roughness, topography, vegetation, buildings, etc.). Near the surface in natural environments, both soil and vegetation exhibit a drag force on the wind field, decreasing wind speeds and creating vertical shear at lower elevations (Holton and

Hakim 2012). The vertical shear induces the development of coherent turbulent structures that play a major role in the characteristics of wind flow within and above canopies (Raupach et al. 1996), as they are responsible for the penetration of high speed wind down into the canopy and stem space from above the canopy and for pushing slower moving air from within the canopy up into the air stream above the canopy. Hence, the canopy plays a critical role in determining the magnitude and fluctuations of wind flows near the surface, and in turn, on fire behavior.

In practice, the most obvious effect of canopy on fire wind is a reduction with respect to a hypothetical wind speed that would be present without the canopy, referred to as "in the open." This effect can be accounted for through Wind Adjustment Factors (WAF) that are used to convert a wind velocity "in the open" to a wind velocity below canopies (Albini and Baughman 1979; Andrews 2012). Over the last three decades, the understanding of the fluid dynamics of turbulent wind flows in canopies has made huge progress thanks to field and wind tunnel experiments (e.g. Raupach et al. 1996) and to modeling (e.g. Dupont and Brunet 2008). In particular, wind profiles can now be characterized from relatively simple stationary models that can be used to estimate WAF (e.g. Massman et al. 2017). Moreover, dynamical flows can be simulated, using Large Eddy Simulations (LES), and exhibit features very similar to observations in the field. The application of such techniques in the context of wildfire behavior started a decade ago (Pimont et al. 2009) and has become increasingly common (Mueller et al. 2014; Frangieh et al. 2018), but the computational cost remains very high and the numerical design is not trivial (Pimont et al. 2020).

Beyond modeling, progress in understanding turbulent flows in plant canopies has also led to a better understanding of how to measure fire wind in field experiments, especially for experimental fires in tree stands. Indeed, it has been shown that the turbulent nature of the flow field is especially challenging in the context of relatively short-duration (~ tens of minutes) and small-scale (~ hundreds of meters) fire experiments, because wind sensors aim at measuring fire wind from safe distances from the fire. The spatial displacement from site of measurement to portions of the fire results in typical random measurement errors on the order of $\pm 30\%$ for usual fire experiments below tree canopy (Sullivan and Knight 2001; Pimont et al. 2017). More recently, it has been shown that canopy heterogeneity, as well as the concept of "wind in the open" tended on the one hand to reduce the magnitude of random measurement errors, but on the other hand to introduce uncontrolled bias of similar magnitude (Pimont et al. 2018).

7.2 Wind Flows in the Lower Part of the Atmosphere

In the lower part of the atmosphere, surface winds are slowed by drag forces resulting from flow interactions with soil and vegetation, and are turned to the left (in Northern Hemisphere), when compared to the geostrophic flow observed well above Earth's surface, which is characterized by a balance between the mesoscale pressure gradient and the Coriolis force. The theoretical balance near the surface between these three forces (pressure, Coriolis, and friction) is called the Ekman balance (Holton and Hakim 2012). It is characterized by a spiral at the top of which the geostrophic wind is parallel to the mesoscale pressure field, whereas the wind direction turns gradually to the left as height decreases in the Northern Hemisphere. The exact magnitude of the rotation depends on the height of the geostrophic layer and the strength of the turbulent mixing, but it is typically on the order of $30°$. It can even be observed within deep and dense canopies, where Coriolis and friction forces, which are proportional to flow velocity, become negligible when compared with the pressure gradient itself (Smith et al. 1972).

In this *planetary boundary layer* (PBL), the surface roughness due to soil and vegetation induces an aerodynamic drag on the winds near the surface. The drag force induced by the vegetation (predominantly by the foliage) is distributed over the height of the vegetation, rather than just on the surface or ground. Heterogeneities in the spatial distribution of the vegetation also induce variability in the flow field. The influence of the potentially heterogeneously distributed drag force on the wind field is often parameterized in terms of a roughness in PBL modeling. This surface roughness influences the mixing within and above the canopy and thus the strength and extent of the vertical gradient in wind velocity in the *surface layer* (SL). This vertical wind shear gives rise to shear – or Kelvin-Helmholtz – instability, responsible for turbulence production, hence coherent structures, which are dissipated by viscous forces, as in any fluid system in which strong velocity gradients are present, such as, for example, a turbulent channel flow (Finnigan 2000; Holton and Hakim 2012).

The combined effects of drag, shear turbulence production, and turbulence dissipation result in a typical logarithmic vertical velocity profile above the ground and the canopy in the *inertial layer* (IL). These patterns can be affected by the vertical profile of potential air temperature,[1] which affects vertical flow motions through buoyancy, and determines whether the atmosphere is stable, neutral, or unstable. With the exception of temperature conditions where the flow is stably stratified, wind flows in the boundary layer are hence turbulent. In addition to the influences of surface features, the near surface winds are also affected by variations of meteorological conditions occurring at synoptic scales.

A general discussion of atmospheric stability, of the impact of topography, and of larger scales is beyond the scope of the present chapter, which focuses on the impact of canopies on wind. It is relevant, however, to analyze the full turbulence

---

[1] The potential temperature of a parcel of fluid at a given pressure is the temperature that the parcel would attain if adiabatically brought to a standard reference pressure.

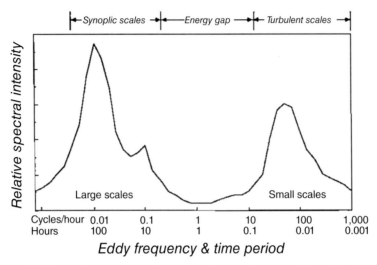

Figure 7.1 Schematic spectrum of wind speed near the ground from Stull (1988). Adapted from Van der Hoven 1957

spectrum of wind speed near the ground (Figure 7.1), in order to better understand the contributions of these different factors near the ground. We can observe that the variation in time of wind velocity mostly occurs at the synoptic scales (a few days), at diurnal scale (secondary peaks on order of 12 hours) and at short times scales (turbulence scale, on the order of a minute). The shear production of turbulence resulting from ground/vegetation effects is one of the main contributions to the high frequency peak "turbulent scales," along with topography and buoyancy. The energy gap between the two large peaks is centered close to 1 hour and corresponds to the minimal variations. There is, as a result, a relative disconnection between variations at large scales (low frequencies), which might steer the fire, and those at fine scales, which influence phenomena that control fire spread. These phenomena include the impact of the vegetation canopy.

7.3 Wind Flows in Homogeneous Canopies

7.3.1 Wind Velocity and Turbulent Statistics Associated with Tree Canopies

The presence of a homogeneous canopy with non-negligible height h modifies the typical logarithmic vertical velocity profile above the canopy in the inertial layer. The canopy height is then a characteristic length of the physical problem. For an elevation z typically higher than 1.5 to 3.5h, the wind profile follows (e.g. Raupach 1994):

$$\bar{U}(z) = \frac{u^*}{\kappa} \ln \left(\frac{z - d}{z_0} \right) \tag{7.1}$$

Parameters in Eq. (7.1) are the roughness length (z_0) and zero-plane displacement (d) of the surface, the friction velocity (u^*) and K ($\sim$0.4) is the von Karman constant. An example time-averaged vertical profile over a plant canopy is shown in Figure 7.2(a). These parameters are affected by the properties of vegetated surfaces, such as the canopy height (h) and Leaf Area Index (*LAI*), as shown in Section 7.5.2. For example, d is typically on the order of ⅔ of h.

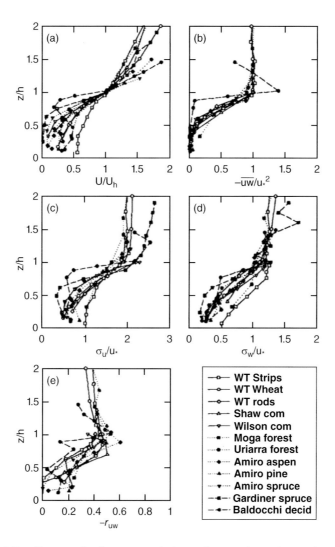

Figure 7.2 Family portrait of canopy turbulence for a variety of canopies reviewed by Raupach et al. (1996), showing profiles with normalized height of normalized (a) U velocity; (b) Momentum flux; (c, d) u and v standard deviations; (e) opposite of the correlation coefficient between u and v velocities; (f, g) u and v skewnesses; and (h, i) u and turbulent length scales.

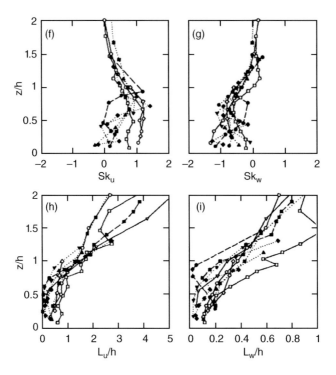

Figure 7.2 (*cont.*)

Below the inertial layer is the *roughness layer* (RL). In this region, the velocity profile exhibits an inflection near the canopy top, which is correlated with the development of large coherent eddies (Kaimal and Finnigan 1994; Raupach et al. 1996). The typical constitutive motions of these structures are strong downdrafts or gusts (referred to as "sweeps") into the canopy and weak updrafts or bursts (referred to as "ejections") vertically out of the top of the canopy. These structures control most of the momentum and scalar transfer between vegetation and the atmosphere (Gao et al. 1989; Lu and Fitzjarrald 1994). Below the inflection point, the wind profile is typically concave (du/dz increases as you go down from the inflection), before merging into the convex profile (du/dz increases as you move upward from the merge point) corresponding to Eq. (7.1). The turbulent kinetic energy and momentum flux also decay quickly within the canopy. A "family" portrait of these main features is presented in Figure 7.2 (from Raupach et al. 1996). In analogy with the plane-mixing layer, it has been suggested to define the shear length scale as follows (Figure 7.2):

$$L_u = \frac{u(h)}{\left. \partial u / \partial z \right|_{z=h}} \qquad (7.2)$$

According to Finnigan (2000), L_u is roughly equal to $0.5h$ and ranges between $0.1h$ and $0.8h$, whether the canopy is dense or light.

7.3.2 Coherent Structures of the Wind Flow in Canopies

The different features of the turbulence in plant canopy are representative of what is called a plane mixing layer, rather than a simple superposition of plant wakes on the surface layer (Raupach et al. 1996; Finnigan 2000). They correspond to large coherent structures that develop in three stages (Figure 7.3): (i) the emergence of primary Kelvin-Helmholtz instability during a shear event, (ii) the clumping into transverse vortices or rollers connected by broad regions of highly strained fluid, and (iii) secondary instabilities in the rollers lead to their kinking and pairing

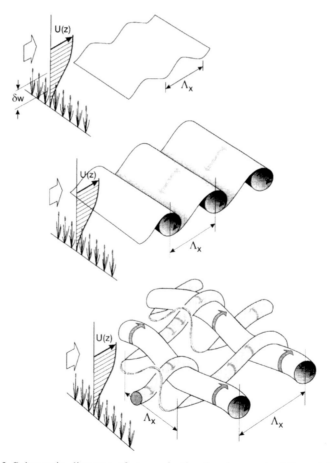

Figure 7.3 Schematic diagram of stages in the development of the mixing-layer type instability in the roughness layer. From Finnigan 2000

(Finnigan 2000). Because 50% of shear stress arises only 10% of the time, the resulting turbulence in plant canopy is highly intermittent. In order to estimate the size of these coherent structures, it has been shown that the two-point correlation technique was the most accurate (Shaw et al. 1995). The typical length scale of spatial correlations is on the order of h in the horizontal and of $1/3h$ in the vertical. The corresponding temporal scales are respectively h/Uh (and $h/(3Uh)$ in the vertical). In the context of tree canopy fires, the vegetation height is typically on the order of 10–20 meters and wind speed on the order of 2–6 ms^{-1}. Hence, the size and duration of coherent structures is typically on the order of 10–20 meters and 2–10 seconds. These large structures typically exhibit a spatial periodicity of $\Lambda_x \approx 8L_u$, which is typically on the order of 50–100 meters for a 20-meter height canopy. These structures can typically be observed in crop canopies, where the wind waves are responsible for coherent plant motions in fields (Dupont et al. 2010).

When vegetation density is lower (smaller LAI), "sweeps" are more frequent and the inflection point is weaker, with a flow that exhibits features intermediate between a surface-layer flow (observed without canopy) and the mixing-layer flow observed in dense canopies (Poggi et al. 2004).

7.3.3 Turbulent Fine Structure

As explained in Sections 7.3.1 and 7.3.2, the turbulence spectrum is strongly affected by the presence of the canopy at small scales (Figure 7.1, less than 0.1 hour). The production of turbulence associated with the shear is responsible for the increase in energy of eddies and the turbulent scale peak (Figure 7.1, between 0.1 and 0.01 hours). The viscous dissipation induces a decrease in energy of eddies with smaller size. This is a general feature of fluid flows, which is known as the Kolmogorov cascade (Figure 7.1, below 0.01 hour). The theory predicts the slope of the decrease in spectral intensity (power –5/3 in the inertial subrange) associated with turbulent structures away from obstruction such as Earth's surface or vegetation (Figure 7.4). A few studies showed that the spectral intensity of wind flows within canopies deviates from this theoretical model, which has been explained by two main reasons. First, the presence of wake production caused by the interactions between the mean wind and foliage elements represents an additional source of small-scale turbulence. Second, the interactions between existing eddies and foliage act as a sink of turbulent kinetic energy for scales larger than canopy elements, which are converted in smaller size eddies. This phenomenon is known as the spectral shortcut (Finnigan 2000) and is illustrated in Figure 7.4.

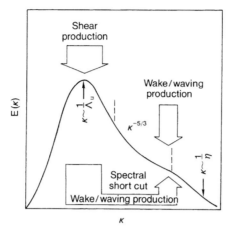

Figure 7.4 Schematic diagram illustration. From Finnigan 2000

7.4 Heterogeneous Canopies

The flow patterns described in Section 7.3 have been extensively studied in dense and homogeneous canopies, on flat terrain and in a neutral atmosphere. One of the main causes of discrepancies in turbulence structure observed in the field is the presence of heterogeneities in plant canopies. The presence of sudden canopy heterogeneities induces additional turbulence, acceleration of the wind, and consequently increased gust intensity (Dupont and Brunet 2008). In the context of wildfires, heterogeneous canopies are of primary importance for two main reasons. First, fire prone ecosystems are often highly heterogeneous, because hetero-geneous patterns in fire severity lead to patchy or aggregated fuel patterns (Turner and Romme 1994). Second, most fuel treatments applied to the tree canopy result in heterogeneous canopies, which can be either small-scale heterogeneities in the case of thinning, or transitions between forest and clearing in the context of clear-cuts, fuel-breaks, or safety breaks.

7.4.1 Tree-Scale Heterogeneity and Sparse Canopies

The changes caused by vegetation structure strongly depend on the characteristic size of the heterogeneities and the cover fraction of the canopy. When the size of the gaps is small (on the order of h) and the cover fraction is high (90%), the wind flow exhibits similar properties as a homogeneous canopy (Patton 1997). As heterogeneity size increases (typically between h and $5h$), wind statistics are modified (Pimont et al. 2011; Parsons et al. 2017). As canopy cover decreases, the wind profiles progressively evolve to open wind profiles (Figure 7.5(a)). When compared to a homogeneous canopy with a same LAI, canopies with larger gaps (denser patches of trees) lead to

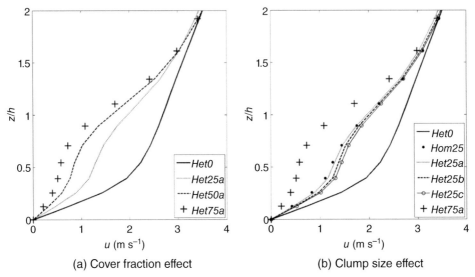

Figure 7.5 Vertical profiles of mean streamwise velocity in the middle of a treated area located in a homogeneous forest. *Het0*, *Het25a*, *Het50a*, and *Het75a* correspond to 2-meter clumps and cover the fraction of, respectively, 0%, 25%, 50%, and 75%. *Het25b* and *Het25c* have a 25% cover fraction, but clumps of 5 and 10 meters. Reproduced from Pimont et al. (2011) with the permission of the Editors in Chief of *Annals of Forest Science*

more heterogenous flows with stronger sweeps leading higher local wind speed in gaps but lower wind speeds within dense patches, compared to a homogeneous canopy. More surprisingly, the resulting mean wind flow (averaging gap and patch wind speeds) is higher when gaps are large than when gaps are small (Figure 7.5(b)). The magnitude of this phenomenon is stronger when tree clumps are denser (Pimont et al. 2011) and larger gaps also increase the variability associated with the specific spatial arrangement of gaps (Parsons et al. 2017).

It is important to have in mind that in the case of a transition between a homogeneous and a heterogeneous forest, the modification of turbulent statistics does not take place just at the location of the transition, contrary to what is assumed in the theoretical fire wind model presented in Andrews (2012). The wind acceleration mainly takes place between $8h$ and $10h$ from the edge of the dense forest block. This is consistent with patterns observed in the case of the more abrupt forest-to-clearing transition, which is described in Section 7.4.3.

7.4.2 Clearing-to-Forest Transition

The clearing-to-forest transition has been extensively studied, including the specific case of wind breaks (Patton et al. 1998). When wind hits the leading edge

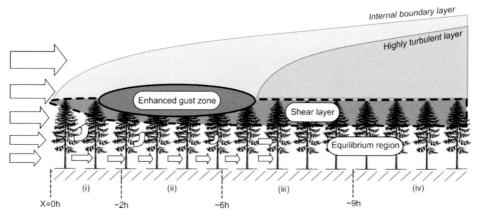

Figure 7.6 Idealized representation of the main characteristics of edge flow over a canopy with an LAI of about 2 and a relatively uniform vertical foliage distribution. From Dupont et al. (2011)

of the canopy, the flow gets distorted with a positive vertical velocity, as the flow is diverted up and over the trees by the positive pressure gradient resulting from the aerodynamic drag of the downwind canopy (Figure 7.6). The canopy drag causes a decrease in horizontal velocity as the flow approaches the forest edge, and the flow accelerates above the canopy (Dupont and Brunet 2008, 2009; Pimont et al. 2009; Dupont et al. 2011). The typical length of adjustment is on the order of $10h$. The large wind shear at the canopy top induces more frequent ejections near the leading edge, which results in the development of a region of sudden strong gusts above the canopy, referred to as the "enhanced gust zone." It typically develops between $2h$ and $6h$ from the leading edge, and plays a critical role in windthrow. It has been shown that canopy density affects the different features of such transitions. In particular, the flow adjusts faster and the enhanced gust zone becomes more distinct when canopy density is high, because the mean upward motion is stronger in this case (Dupont and Brunet 2008). This enhanced gust zone should probably be considered in fire experiments, as experimental fire plots are typically located downwind from a clearing, within an ignition line along this edge. Hence, the initial spread of the fire typically occurs in this zone, which is not necessarily representative of the "homogenous canopy" flow in terms of wind flow.

7.4.3 *Forest-to-Clearing Transition*

The transition between forest and clearing corresponds to the typical configuration of a fire spreading from a forest to a safety break. In this case the wind flow accelerates because of the decrease in drag force. This acceleration is associated with a mean downward motion, resulting in a negative vertical flux of higher-velocity

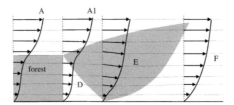

Figure 7.7 Schematic diagram of exit flow across a two-dimensional forest edge. The flow field in the open is divided into quiet zone (D), mixing zone (E), and re-equilibration zone (F). From Lee 2000

winds in the upwind portion of the fuel-break (Pimont et al. 2009, 2011). According to Lee (2000), three different zones can be identified (Figure 7.7). The quiet zone between the edge and $4h$–$7h$, in which the wind profile remains close to the one of the forest zone, is followed by the mixing zone, in which mixing is enhanced near the canopy top height at roughly $5h$ from the edge, with intermittent recirculating eddies. In this region, the large amount of turbulent kinetic energy located near the tree top (as seen in Figure 7.2(c) for example) is advected downwind from the tree canopy. The last zone corresponds to the re-equilibration zone, which only takes place beyond $22h$. Recent studies suggest that more than $30h$ can be required for a full adjustment of the profile (Pimont et al. 2018). These values are typically much higher than what is usually recommended for measuring the wind velocity "in the open" (e.g. $7h$ in Fischer and Hardy 1976), which raises the question of the representativeness of such measurements. This question is addressed with more details in Section 7.6.

7.5 Modeling the Interaction between Wind Flow and Plant Canopies

Wildfire models range from the empirical to conglomerations of theoretical representations of physical processes that drive wildfires. Winds are incorporated into these models in different ways depending on the basis behind the model development. In empirical models, the mean fire wind at specific heights is used. It can be the average wind speed at mid-flame height (see, for example, the BEHAVE model in Andrews 1986), or the 10-meter-height wind speed measured from meteorological stations. The appropriate reference wind speed is not easy to define (Linn and Cunningham 2005) and is influenced by the local environment, including fire and fuel discontinuities. The modifications of wind–field dynamics and turbulence regimes due to fuel discontinuities are not often explicitly considered (some counter examples are Linn et al. 2012; Pimont et al. 2014; and Dupuy et al. 2014). Additionally, the empirical laws of fire used by some of these models are deduced from laboratory experiments (Rothermel 1972; Catchpole et al. 1998), where turbulence conditions can be very different from turbulence generated in field canopies. In recent physics-based fire behavior models, the physical processes driving wildfires include coupled atmosphere–fire interaction and wind interaction with canopies. In these models, the determination of fire behavior (spread, intensity, etc.) depends on physical processes occurring within and above the vegetation.

The physics of motion of air, which is a viscous fluid, are described by the Navier Stokes equations. These equations state that the rate of change of the momentum is proportional to the divergence of the stress tensor, which is equal to the sum of a diffusive viscous term and a pressure term (Pope 2000). The solution of the equations is the flow velocity field, defined at any moment in a time interval. In general, flow fields can be either laminar or turbulent, but most wind flows within and above canopies are turbulent, as shown in Section 7.3. Although a technique called DNS (Pope 2000) endeavors to reproduce the very detailed motion of fluid parcels, practical modeling approaches average motions both in time and space and represent the contribution of small-scale motion to turbulence through a quantity called the turbulent kinetic energy, which is one-half the variance of the wind velocity magnitude. In practice, it is often only the mean wind velocity that is considered as an input factor in many fire models.

7.5.1 The Navier-Stokes Equations and the Drag Force

Mass equation

$$\frac{\partial \rho}{\partial t} + \nabla(\rho u) = 0. \tag{7.3}$$

Momentum equation

$$\frac{\partial u}{\partial t} + u \nabla u = -\frac{1}{\rho} \nabla p + v \nabla^2 u. \tag{7.4}$$

Scalar equation

$$\frac{\partial \theta}{\partial t} + u \nabla \theta = \frac{v}{Pr} \nabla^2 \theta. \tag{7.5}$$

In models that represent the structure of the vegetation, the impact of vegetation on the mean velocity is usually represented in terms of drag force, added to the right-hand side of the momentum equation (Eq. (7.4)). This type of formulation accounts for both viscous and pressure drags and is a function of Leaf Area Density (LAD, m^2 m^{-3}), which expresses the amount of leaf area per unit volume. This quantity is similar to the notion of bulk density (kg m^{-3}), which is more familiar to fire researchers, but for leaf area instead of for leaf mass. The integration of LAD over the vertical leads to the well-known LAI, in a similar manner that the load (kg m^{-2}) is the vertical integration of the bulk density. The drag force in a particular direction is also a function of wind velocity in that direction and of a parameter c, which theoretically varies with the Reynolds number. A review of several experimental and modeling studies in Pimont (2008) showed that a typical value for C ranged between 0.1 and 0.3, which is also suggested in Massman et al. (2017):

$$F = cLAD|u|u. \qquad (7.6)$$

For the turbulent flows in plant canopies, the Navier-Stokes equations do not exhibit any analytical solution and require significant numerical and computational resources to be solved. For this reason, some simple analytical models have been developed based on a mixture of theoretical support and empirical data. These models are described in Section 7.5.2. Methods and models used to numerically solve the Navier-Stokes equations are briefly described in Section 7.5.3 onwards.

7.5.2 One-Dimensional Analytical Models

This simple class of models aims at simulating the mean wind profile within and above a canopy, from a combination of theoretical knowledge and field data (e.g. Cionco 1965; Raupach 1994; Massman et al. 2017). Below the canopy, the profile is typically exponential, whereas the modified log profile of the inertial layer (Eq. (7.1)) is used high above the canopy. An example of such a simple parametrization, which is function of LAI and U_h (the wind velocity at canopy top), is summarized in Cassagne et al. (2011) and is presented here with simplified notations and the sign correction of Raupach (1995):

$$\text{if } z \le h, u(z) = U_h \exp\left(-a_1\left(1 - \frac{z}{h}\right)\right), \text{with } a_1 = \min\left(0.5\text{LAI} + 1.2; 3.2\right)$$

$$\text{if } z \ge 2h, u(z) = a_2 U_h \log\left(\frac{\frac{z}{h} - 1 + a_3}{a_4}\right),$$

$$\text{with } a_2 = \min\left(\sqrt{0.018 + 0.89\text{LAI}}; 0.73\right),$$

$$a_3 = \frac{1 - \exp\left(-\sqrt{7.5\text{LAI}}\right)}{\sqrt{7.5\text{LAI}}} \text{ and } a_4 = a_3 \exp\left(-\frac{1}{a_2} + 0.19\right)$$

$$\text{if } h < z < 2h, u(z) = \left(2 - \frac{z}{h}\right)u(h) + \left(\frac{z}{h} - 1\right)u(2h). \qquad (7.7)$$

However, such a parametrization is only approximative, since the LAD profile and the heterogeneity in fuel structure strongly affect this mean profile. A more sophisticated model was recently adapted in the context of fire modeling by Massman et al. (2017). It includes the vertical distribution of the LAD profile (not only the LAI), which is represented through an asymmetric gaussian model for simplicity. Such a model uses a product of logarithm and hyperbolic functions to improve the shape of the wind profile below the canopy. As for any one-dimensional analytical model, it exhibits two limitations: first, it does not account for horizontal vegetation heterogeneity. Second, it neglects the variations in wind profile shapes that are caused by variations in wind speed magnitude above the canopy, since wind speed at a given height z is proportional to wind velocity at canopy top U_h (as in Eq. (7.7)). A recent study based on field data found that the shape of such profiles was highly variable with wind speed magnitude, especially when open wind speed was lower than 8–12 km h^{-1} (Moon et al. 2019). Because of this effect, the corresponding change in wind adjustment factor could vary by a factor greater than four. More sophisticated CFD-type models that incorporate more physics of fluid motion can overcome these two limitations (e.g. Pimont et al. 2009), but they require significantly more computations.

7.5.3 The Reynolds-Averaged Navier-Stokes Approach

Because of the turbulent nature of the flow, laminar solvers fail to correctly represent the Navier-Stokes equations. One solution is to average the equations in time in order to simplify the study of the fluctuating quantities. The Reynolds-Averaged Navier-Stokes equations (RANS) relies on a time average of instantaneous quantities in the case of stationary flows. However, the averaging operation leads to the appearance of stress terms in the momentum equations, called Reynold stress, which represent the average aggregated impacts of the temporal fluctuations in the velocity field. The modeling of these Reynold stress terms requires additional assumptions to close and solve the RANS equations. This approach is typically steady-state, as a consequence of time averaging, and is well-suited for the representation of a steady state flow in plant canopies (Li et al. 1990; Liu et al. 1996; Foudhil et al. 2005). The turbulence closure relies on the computation of a turbulence length scale representing the size of dominant turbulent structures. The value for such length scales can be either fixed by the user (Li et al. 1990) or computed with either K-s models (where the turbulent kinetic energy, K, turbulence length scale, s, and the turbulence dissipation, ε are related by the equation $s = K^{3/2}/\varepsilon$) or $K - \varepsilon$ models, which uses an additional equation for dissipation (Liu et al. 1996; Foudhil et al. 2005). These models compute the mean flow (and hence the wind velocity profile), and the turbulent kinetic energy, which summarizes the energy associated with the flow deviations from the mean flow (the fluctuations in the flow). In these models, the effect of the vegetation drag is represented by the usual drag force, which acts on the mean flow, but they can also implement the wake production and spectral short cuts (see Section 7.3.3), which represent the interaction between wind fluctuations and vegetation elements in these models (e.g. Liu et al. 1996; Foudhil et al. 2005). In the context of wildfire modeling, a typical example of a $K - \varepsilon$ model is FIRESTAR (Morvan and Dupuy 2001). These models have been improved in their representation of turbulence, with the modeling of higher-order statistics, leading to more complex closure and additional equations, but they did not always succeed in correctly representing these high order statistics (Katul 1998). Another limitation of these models is that they only simulate the mean flow and turbulent statistics, but they do not explicitly simulate the turbulent and intermittent structures and fluctuations, which are of primary interest at canopy and fire scales. In the context of forest fire, the meaning of temporal averages is also not obvious, because of the spread of the fire inside the domain, leading to the concept of "Unsteady" RANS (U-RANS).

7.5.4 *Large Eddy Simulation*

An intermediate approach between the direct numerical simulation (DNS), which aims to model the whole spectrum of the turbulence, and the RANS approach, which fully average fluctuations, is the large eddy simulation (LES). This approach intends to explicitly represent turbulent flow patterns whose length scales are large enough to resolve and use a RANS-style approach to represent the energy associated with the finer-scale fluctuations. The philosophy is different for LES than RANS, since the averaging is spatial rather than temporal, so that the models solve all fluctuations larger than this spatial filter and simulates the evolution of local velocity fields over time. This approach requires more computational resources, since it has to be run in a transient mode, but it permits the explicit simulation of coherent structures with size at least twice as large as the local grid cells. The turbulence energy associated with smaller structures is estimated using a subgrid turbulent kinetic energy model. The formulations of these subgrid turbulence kinetic energy models can take a variety of forms ranging from local diagnostic shear-based terms to separate transport equations, which are relatively similar to the K equation in a RANS approach. Initially applied with grid cell much larger than canopy height in meteorological studies (Deardorff 1980), LES has been applied with success to canopy flows, with a grid size typically on the order of 2 meters, in canopies higher than 10 meters. Numerical studies deal with both homogeneous (Shaw and Schumann 1992; Su et al. 1998; Dupont et al. 2011; Pimont et al. 2017) and heterogeneous (Patton et al. 1998; Dupont and Brunet

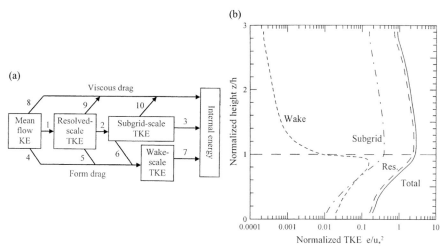

Figure 7.8 Schematic of the processes that convert resolved-scale kinetic energy to subgrid-scale, wake scale, and ultimately to internal energy. (a) Flow-chart indicating the direction of energy transfer between scales, and (b) vertical structure of turbulent kinetic energy across the canopy, showing the division of energy into the respective components and relative strength. From Shaw and Patton (2003)

2008) vegetation canopies under essentially neutral stratification. As for the RANS model, the wake production and spectral shortcut can be implemented in LES equations (Shaw and Patton 2003; Dupont and Brunet 2008, 2009; Pimont et al. 2009), as shown in Figure 7.8. This technique has been applied to a wide variety of investigations. For instance, these simulations have been analyzed in detail to understand the mechanisms associated with interactions between canopy and wind flow (Shaw and Patton 2003; Dupont and Brunet 2008, 2009; Dupont et al. 2010) and to determine the spatial extent and other important properties of transition zones in heterogeneous canopies (Dupont and Brunet 2008, 2009; Pimont et al. 2011, Dupont et al. 2011).

In the context of fire modeling, LES has been mostly used with the FIRETEC model, following the work described in Pimont et al. (2009), but also occasionally with WFDS (Mueller et al. 2014) and FIRESTAR (Frangieh et al. 2018). The Arps-canopy model, which has been extensively used in LES of canopy flows (e.g. Dupont and Brunet 2008), has also been applied to several fire studies (e.g. Kiefer et al. 2013). However, horizontal resolutions for fire applications were generally larger than 30 meters, so that the model runs did not explicitly simulate the coherent structures generated by canopy shear, which would require a higher horizontal resolution on the order of 2 meters, as in Dupont and Brunet (2008).

It is important to acknowledge that the design of numerical simulations is critical to accurately simulate realistic canopy flows with physics-based models. Beyond resolution, another critical aspect is the use of cyclic boundary conditions. Indeed, it has been shown that "static" simulations based on specified static inflow conditions (such as a constant wind profile) failed to reproduce correctly the turbulent statistics without the use of excessively long domains in the windward direction (Pimont 2008; Mueller et al. 2014). Numerical investigations done with FIRETEC on a very long domain showed that such structures typically begin to develop from a static inlet beyond 1,000 meters from the boundary, which is beyond the size of most simulations done with physically based models. The requirement for very long domains can be overcome by using cyclic boundary conditions in simulations where winds exit the domain on the downstream side and re-enter on the upstream side, emulating aspects of an infinitely long domain. This approach has proven to be a more efficient way of developing realistic turbulent motions, even if the same equations and solvers were used in both cases. This result might seem surprising, but the explanation is quite straightforward: the large coherent structures in canopy flow are a result of the combining of other precursor flow patterns and thus do not develop instantaneously (see for example the development of the coherent structures described in Figure 7.3 or the large distances involved in transitions in the context of heterogeneous canopies). Hence, it is not surprising that the turbulence arising from a given inlet profile does not lead to explicit coherent structures at a distance from the inlet on the order of a few times the height of the canopy h. Using cyclic boundary conditions provide a means by which to use much smaller domains to simulate realistic wind flows. It is worth acknowledging that the cyclic domain approach does not allow one to shorten the time for spin-up of the turbulent structure as the simulations still need to be run for long enough to emulate the flow over very long distances (turbulence takes on the order of 20 minutes of simulated time to settle in a run done in a domain of a few hundred meters long).

An additional complicating factor for the use of cyclic conditions for the development of realistic winds is the fact that the net impact of vegetation drag force acts to oppose the mean flow. It is thus a net sink of momentum and eventually slows the flow as the default implementation of the cyclic boundaries does not provide any way to replenish this momentum. In reality this momentum is replenished by large-scale pressure forces. Without this pressure force and the effects of the Coriolis force the cyclic boundary condition simulation also tends to generate large persistent unrealistic structures. Pimont et al. (2020) have shown that such unrealistic features disappeared when a forcing based on a mesoscale pressure gradient and the Coriolis force was applied (a numerical design corresponding to the theoretical Ekman balance described in Section 7.1), because of the stability[2] of

---

[2] Stability, in the sense that a small perturbation from the equilibrium is damped by the combination of the forces.

the Ekman balance. With the Ekman balance, these "streaks" are transient and of much smaller magnitude, and are, in this sense, realistic. However, this numerical design raises practical problems in the context of wildfire simulations. In particular, it is difficult to specify specific desired wind speed and direction at a reference height, and rotation of the flow near the canopy is neither controllable nor fully realistic. A new technique based on a specific large-scale pressure gradient forcing has been developed in Pimont et al. (2020) to overcome these different limitations, and provide satisfactory simulations of wind flows in the context of wildfire simulations. Once realistic wind flows are computed through a method such as the cyclic domain technique, the dynamic wind flow can be used to provide initial and boundary conditions in a fire simulation (e.g. Cassagne et al. 2011). This technique has been used in most FIRETEC studies including tree canopies after 2013.

Beyond the differences associated with the method used to represent the impacts of the velocity fluctuations in the solution of the Navier-Stokes equations (RANS, LES) and with the numerical design, it is important to acknowledge that there is a much wider consensus on the formulation of the models in this research field (equation, drag formulation, turbulence, etc.), than in the field of forest fire modeling. In general, models for wind flows are more easily evaluated against field data than fire behavior models, since the quantity of interest is well-defined and can be measured directly over long periods of time. To date, only a limited number of fire models have been evaluated in their capability to simulate wind flows. Moreover, some of them are not run with a design that is expected to lead to realistic turbulent features. This opens the way for significant future improvements in fire modeling.

7.5.5 Measuring and Modeling Canopy Properties

Sections 7.2–7.4 showed that a critical factor affecting the wind flow was the three-dimensional LAD of the tree canopy, which controls the spatial distribution of vegetation drag. LAD is related to the bulk density distribution of leaves through the Leaf to Mass Area coefficient. Physics-based models can account for such detailed information and software packages have been developed to generate realistic three-dimensionally resolved LAD distributions associated with a wide range of forests and woodlands, such as FuelManager (Pimont et al. 2016) or STANDFIRE (Parsons et al. 2018). However, key parameters of the LAD distribution control most of the impact of the vegetation on wind flow. The most important parameter is canopy height h (the height above which LAD is zero). The canopy height plays and important role in setting flow structure length scales, distance of influence, etc. The second critical parameter is the LAI, which is the vertical integration of the LAD. LAI controls the overall amount of drag and has a strong impact on mean flow (e.g. Eq. (7.7)). However, as pointed out in Sections 7.3 and 7.4, the wind profile and other turbulent statistics are strongly affected by the shape of the LAD profile, as well as the presence of heterogeneities. For a detailed study of wind flow, it is therefore critical to account for the three-dimensional distribution of the LAD field. To date, realistic LAD distribution in model domains is based on field measurements such as inventory-based approach or statistical models, but also on some promising terrestrial LiDAR measurements (e.g. Béland et al. 2011; Pimont et al. 2015).

7.6 Measuring Wind in Fire Experiments

Fire experiments generally aim to relate fire behavior to fuel and weather conditions, especially wind speed. However, empirical relationships generally exhibit significant variability, which can partly be explained by wind measurement accuracy (Cruz and Alexander 2013). Recent studies showed that data accuracy was critical in the context of fire modeling (Linn et al. 2012; Pimont et al. 2019). In particular, measuring the wind in fire experiments is challenging, due to the turbulent nature of wind flow in canopies and the remote location of sensor(s). To a large degree, the question of wind measurement representativeness, although critical, has seldom been addressed. Beyond modeling, the progress in understanding of turbulent flows in plant canopies described in Sections 7.3 and 7.4 has also led to a better understanding of how to measure fire wind. Indeed, because the turbulent nature of wind flow in the canopy is dominated by large and intermittent turbulent structures, it is now well acknowledged that there is very limited correlation between the instantaneous local wind velocities measured by remote sensors located on the wind tower (measured wind) and the instantaneous fire wind, which are local to the fireline (within distance often on the order of several times the height of the canopy (Sullivan and Knight 2001; Taylor et al. 2004; Pimont et al. 2017). Data suggest that this low correlation occurs even when considering a lag between the winds at these two locations (Taylor et al. 2004). This is all the more important as Linn et al. (2012) showed that variations in time series of wind speed strongly affect the fire behavior. The correlation – and in turn the meaning of the measured data – only arises from the temporal and spatial averaging of both measured and fire winds.

The temporal average is related to the duration of the fire experiment, whereas the spatial average is related to the length of the fire front (hence fire plot size) and number of sensors, located some distance apart. Both Sullivan and Knight (2001) from field measurements and Pimont et al. (2017) from LES showed that typical random measurement errors are on the order of $\pm 30\%$ for usual fire experiments in which wind is measured below the tree canopy. Such errors can be reduced by

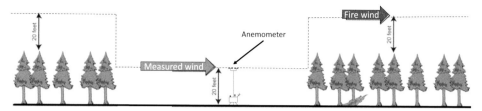

Figure 7.9 Schematic view of the concept of measured and fire wind corresponding to a remote measurement of wind velocity during fire experiments. Adapted from Andrews 2012

increasing the duration of the experiment, the size of the experimental plot, and the number of wind sensors. However, one of the strong limitations of these two studies were that both measured and fire wind were evaluated below homogeneous canopies. In practice, the wind is measured "in the open" and a widespread assumption is that the mean wind at $h_{ref} = 20$ feet (or 10 meters) above the ground is similar to the wind at $h_{ref} = 20$ feet (or 10 meters) above the canopy (Figure 7.9, from Andrews 2012). However, there is apparently no significant scientific support for this assumption. According to Fischer and Hardy (1976), measurement in an open area can be replaced by measurement at "obstacle + h_{ref}" height; "if the ground is densely covered with rocks, brush, or small trees, increase the height of the anemometer by the average height of the ground cover." A weaker assumption would be that the wind in the open is representative of the ambient fire wind, without exact consideration of height and location.

Pimont et al. (2018) showed from a LES study that measuring wind in the open instead of below canopy significantly reduced the magnitude of these random errors, since the wind flow is far less variable above the ground or canopies than within canopies (e.g. Pimont et al. 2009). However, this study also highlighted (see Figure 7.10) that a significant bias may exist between mean values of wind speed in the open and above the canopy at fire location. The magnitude of such bias was shown to be highly sensitive to the location of the sensor in the fuel break, and in particular to the distance d from the trailing edge of the upwind forest block. Indeed, the wind profile above the canopy is not just a vertical "shift" of the wind profile compared to the wind in the fuel break, as shown in Sections 7.3 and 7.4. Moreover, the adjustment distance of the wind profile in the context of the forest to clearing transition is larger than $20h$, so that in order to avoid measuring artifacts of the canopy-to-open transition, the wind tower should be located at more than 400 meters from the trailing edge for a 20 meter canopy height, which is not feasible in many contexts. According to Pimont et al. (2018), a distance to clearing of $7h$ as recommended in Fischer and Hardy (1976) and Andrews (2012) is clearly insufficient for such an adjustment.

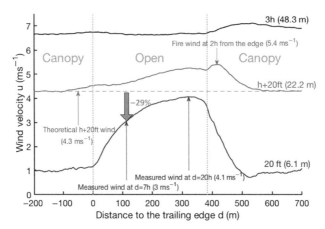

Figure 7.10 Horizontal wind profiles corresponding to Figure 7.9 at various heights. Bottom line: the wind velocity at sensor heights (20 ft), which exhibits a strong increase along the open until a distance of $20h$ to the trailing edge of the first forest block. Medium line: the wind velocity at $h + 20$ ft and in black the velocity at $3h$.

A black and white version of this figure will appear in some formats. For the color version, please refer to the plate section.

Because the locations of wind sensors – and in particular the distance from the upwind forest edge – are often unknown and generally highly variable within a set of fire experiments, Pimont et al. (2018) showed that the corresponding variation in bias between wind and fire wind acted as a secondary source of errors in fire wind measurement (referred to as "lack of control"). For example, pulling wind data measured at various distances between 5 and $20h$ could lead to a MAPE between fire wind and measured wind of more than 19%. They, hence, recommend recording the distances from the wind measurement to the upwind forest, canopy height of the upstream forest when taking wind measurements in fire experiments. These parameter will help in the development of "horizontal wind adjustment factor" (in a similar manner to the WAF for fire wind height), in future fire experiments to help control and correct the bias resulting from open wind measurement location, and thus reduce the measurement error of fire wind.

7.7 Conclusion

Wind is one of the key factors affecting wildfires and the nature of the turbulent flow around fire is strongly influenced by the presence of tree canopies. Fluid dynamics science reveals that the canopy flows exhibit specific features including typical profiles with an inflection point, a large amount of turbulent kinetic energy, and short-duration intermittent coherent structures of a size on the order of the canopy height. These different features, often acknowledged in practice as gusts, are likely to strongly affect fire behavior, and thus wind in the canopy should not be considered as a simple constant laminar flow.

A large variety of models are available to describe wind in the canopy, from one-dimensional analytical models to very complex LES models. All of which can be used to compute wind adjustment factors, which are critical to convert ambient wind to fire wind. The more detailed and complex LES approaches can explicitly simulate coherent structures of a few meters, provided that the numerical design is appropriate. Unfortunately, this numerical design requires special care in setting up simulations, which is often ignored in the case of fire simulations. Insights from fluid dynamic science and of detailed LES models are useful to improve our understanding of wind in the context of fire behavior and fire experiments, and in particular to quantify, interpret, and potentially reduce measurement errors associated with the remote measurement of fire wind. However, simulating the details of wind flow turbulence comes with a high degree of sophistication in the model and requires very significant computational means, which are not necessary required for all fire behavior modeling applications.

References

Albini, FA (1985) A model for fire spread in wildland fuels by-radiation. *Combustion Science and Technology* **42**(5–6), 229–258.

Albini, FA, Baughman, RG (1979) Intermountain Forest and Range Experiment Station (Ogden, Utah) (1979) "Estimating windspeeds for predicting wildland fire behavior." Research report, Intermountain Forest and Range Experiment Station, Forest Service, U.S. Dept. of Agriculture: Ogden, UT.

Andrews, PL (1986) BEHAVE: Fire behavior prediction and fuel modeling system-BURN Subsystem, part 1 (No. INT-GTR-194). Research Report, U.S. Department of Agriculture, Forest Service, Intermountain Research Station, Ogden, UT.

Andrews, PL (2012) *Modeling wind adjustment factor and midflame wind speed for Rothermel's surface fire spread model.* Ft. Collins, CO: United States Department of Agriculture/Forest Service, Rocky Mountain Research Station.

Baines, PG (1990) Physical mechanisms for the propagation of surface fires. *Mathematical and Computer Modelling* **13**(12), 83–94.

Béland, M, Widlowski, J-L, Fournier, RA, Côté, J-F, Verstraete, MM (2011) Estimating leaf area distribution in savanna trees from terrestrial LiDAR measurements. *Agricultural and Forest Meteorology* **151**(9), 1252–1266.

Cassagne, N, Pimont, F, Dupuy, J-L, Linn, RR, Mårell, A, Oliveri, C, Rigolot, E (2011) Using a fire propagation model to assess the efficiency of prescribed burning in reducing the fire hazard. *Ecological Modelling* **222**(8), 1502–1514.

Catchpole, WR, Catchpole, EA, Butler, BW, Rothermel, RC, Morris, GA, Latham, DJ (1998) Rate of spread of free-burning fires in woody fuels in a wind tunnel. *Combustion Science and Technology* **131**(1–6), 1–37.

Cheney, NP, Gould, JS, Catchpole, WR (1998) Prediction of fire spread in grasslands. *International Journal of Wildland Fire* **8**(1), 1–13.

Cionco, RM (1965) A mathematical model for air flow in a vegetative canopy. *Journal of Applied Meteorology* **4**, 517–522.

Cruz, MG, Alexander, ME (2013) Uncertainty associated with model predictions of surface and crown fire rates of spread. *Environmental Modelling & Software* **47**, 16–28.

Cruz, MG, Alexander, ME, Wakimoto, RH (2005) Development and testing of models for predicting crown fire rate of spread in conifer forest stands. *Canadian Journal of Forest Research* **35**(7), 1626–1639.

Deardorff, JW (1980) Stratocumulus-capped mixed layers derived from a three-dimensional model. *Boundary-Layer Meteorology* **18**(4), 495–527.

Dupont, S, Bonnefond, J-M, Irvine, MR, Lamaud, E, Brunet, Y (2011) Long-distance edge effects in a pine forest with a deep and sparse trunk space: In situ and numerical experiments. *Agricultural and Forest Meteorology* **151**(3), 328–344.

Dupont, S, Brunet, Y (2008) Edge flow and canopy structure: A large-eddy simulation study. *Boundary-Layer Meteorology* **126**(1), 51–71.

Dupont, S, Brunet, Y (2009) Coherent structures in canopy edge flow: A large-eddy simulation study. *Journal of Fluid Mechanics* **630**, 93–128.

Dupont, S, Gosselin, F, Py, C, De Langre, E, Hemon, P, Brunet, Y (2010) Modelling waving crops using large-eddy simulation: comparison with experiments and a linear stability analysis. *Journal of Fluid Mechanics* **652**, 5–44.

Dupuy, J-L, Pimont, F, Linn, R, Clements, C (2014) FIRETEC evaluation against the FireFlux experiment: preliminary results. In: Viegas, DX, ed., *Advances in Forest Fire Research*. Coimbra: Imprensa da Universidade de Coimbra, pp. 261–274.

Finnigan, J (2000) Turbulence in plant canopies. *Annual Review of Fluid Mechanics* **32**(1), 519–571.

Fischer, WC, Hardy, CE (1976) Fire–weather observers' handbook. *Agricultural Handbook*, vol. 494. Washington, DC: USDA Forest Service.

Foudhil, H, Brunet, Y, Caltagirone, JP (2005) A fine-scale k-e model for atmospheric flow over heterogeneous landscapes. *Environmental Fluid Mechanics* **5**(3), 247–265.

Frangieh, N, Morvan, D, Meradji, S, Accary, G, Bessonov, O (2018) Numerical simulation of grassland fires behavior using an implicit physical multiphase model. *Fire Safety Journal* **102**, 37–47.

Gao, W, Shaw, RH, Paw, UKT (1989) Observation of organised structures in turbulent flow within and above a forest canopy. *Boundary-Layer Meteorology* **47**(1), 349–377.

Hernandez, C, Keribin, C, Drobinski, P, Turquety, S (2015) Statistical modelling of wildfire size and intensity: A step toward meteorological forecasting of summer extreme fire risk. *Annales Geophysicae* **33**(12), 1495–1506.

Holton, JR, Hakim, GJ (2012) *An Introduction to Dynamic Meteorology*, 5th ed. Amsterdam: Academic Press.

Kaimal, JC, Finnigan, JJ (1994) *Atmospheric Boundary Layer Flows: Their Structure and Measurement*. New York: Oxford University Press.

Katul, G (1998) An investigation of higher-order closure models for a forested canopy. *Boundary-Layer Meteorology* **89**(1), 47–74.

Kiefer, MT, Zhong, S, Heilman, WE, Charney, JJ, Bian, X (2013) Evaluation of an ARPS-based canopy flow modeling system for use in future operational smoke prediction efforts: ARPS-CANOPY EVALUATION. *Journal of Geophysical Research: Atmospheres* **118**(12), 6175–6188.

Lee, X (2000) Air motion within and above forest vegetation in non-ideal conditions. *Forest Ecology and Management* **135**(1–3), 3–18.

Li, Z, Lin, JD, Miller, DR (1990) Air flow over and through a forest edge: A steady state numerical simulation. *Boundary Layer and Meteorology* **51**(1), 179–197.

Linn, R, Anderson, K, Winterkamp, J, Brooks, A, Wotton, M, Dupuy, J-L, Pimont, F, Edminster, C (2012) Incorporating field wind data into FIRETEC simulations of the International Crown Fire Modeling Experiment (ICFME): Preliminary lessons learned. *Canadian Journal of Forest Research* **42**(5), 879–898.

Linn, RR, Cunningham, P (2005) Numerical simulations of grass fires using a coupled atmosphere–fire model: Basic fire behavior and dependence on wind speed. *Journal of Geophysical Research* **110**(D13), D13107.

Liu, J, Chen, JM, Black, TA, Novak, MD (1996) E-e modelling of turbulent air flow downwind of a model forest edge. *Boundary-Layer Meteorology* **77**(1), 21–44.

Lu, C-H, Fitzjarrald, DR (1994) Seasonal and diurnal variations of coherent structures over a deciduous forest. *Boundary-Layer Meteorology* **69**(1–2), 43–69.

Massman, WJ, Forthofer, JM, Finney, MA (2017) An improved canopy wind model for predicting wind adjustment factors and wildland fire behavior. *Canadian Journal of Forest Research* **47**(5), 594–603.

Moon, K, Duff, TJ, Tolhurst, KG (2019) Sub-canopy forest winds: understanding wind profiles for fire behaviour simulation. *Fire Safety Journal* **105**, 320–329.

Morvan, D, Dupuy, JL (2001) Modeling of fire spread through a forest fuel bed using a multiphase formulation. *Combustion and Flame* **127**(1), 1981–1994.

Mueller, E, Mell, W, Simeoni, A (2014) Large eddy simulation of forest canopy flow for wildland fire modeling. *Canadian Journal of Forest Research* **44**(12), 1534–1544.

Parsons, RA, Linn, RR, Pimont, F, Hoffman, C, Sauer, J, Winterkamp, J, Sieg, CH, Jolly, WM (2017) Numerical investigation of aggregated fuel spatial pattern impacts on fire behavior. *Land* **6**(2), 43.

Parsons, RA, Pimont, F, Wells, L, Cohn, G, Jolly, WM, de Coligny, F, Rigolot, E, Dupuy, J-L, Mell, W, Linn, RR (2018) Modeling thinning effects on fire behavior with STANDFIRE. *Annals of Forest Science* **75**(1), 1–10.

Patton, EG (1997) *Large-Eddy Simulation of Turbulent Flow above and within a Plant Canopy*. PhD Dissertation. University of California, Davis, CA.

Patton, EG, Shaw, RH, Judd, MJ, Raupach, MR (1998) Large-eddy simulation of wind-break flow. *Boundary-Layer Meteorology* **87**(2), 275–307.

Pimont, F (2008) *Modélisation physique de la propagation des feux de forêts: effets des caractéristiques physiques du combustible et de son hétérogénéité*. PhD Dissertation. Université d'Aix-Marseille II.

Pimont, F, Dupuy, J-L, Linn, RR (2014) Fire effects on the physical environment in the WUI using FIRETEC. In: Viegas, DX, ed., *Advances in Forest Fire Research*. Coimbra: Imprensa da Universidade de Coimbra, pp. 749–758.

Pimont, F, Dupuy, J-L, Linn, RR, Dupont, S (2009) Validation of FIRETEC wind-flows over a canopy and a fuel-break. *International Journal of Wildland Fire* **18**(7), 775.

Pimont, F, Dupuy, J-L, Linn, RR, Dupont, S (2011) Impacts of tree canopy structure on wind flows and fire propagation simulated with FIRETEC. *Annals of Forest Science* **68**(3), 523–530.

Pimont, F, Dupuy, J-L, Linn, RR, Parsons, R (2018) Wind measurement accuracy in fire experiments. In: Viegas, DX, ed., *Advances in Forest Fire Research*. Coimbra: Imprensa da Universidade de Coimbra, pp. 716–724.

Pimont, F, Dupuy, J-L, Linn, RR, Parsons, R, Martin-StPaul, N (2017) Representativeness of wind measurements in fire experiments: Lessons learned from large-eddy simulations in a homogeneous forest. *Agricultural and Forest Meteorology* **232**, 479–488.

Pimont F, Dupuy J-L, Linn RR, Sauer JA, Muñoz-Esparza D (2020) Pressure-gradient forcing methods for large-eddy simulations of flows in the lower atmospheric boundary layer. *Atmosphere* **11**(12), 1343.

Pimont, F, Dupuy, J-L, Rigolot, E, Prat, V, Piboule, A (2015) Estimating leaf bulk density distribution in a tree canopy using terrestrial LiDAR and a straightforward calibration procedure. *Remote Sensing* **7**(6), 7995–8018.

Pimont, F, Parsons, R, Rigolot, E, de Coligny, F, Dupuy, J-L, Dreyfus, P, Linn, RR (2016) Modeling fuels and fire effects in 3D: Model description and applications. *Environmental Modelling & Software* **80**, 225–244.

Pimont, F, Ruffault, J, Martin-StPaul, NK, Dupuy, J-L (2019) Why is the effect of live fuel moisture content on fire rate of spread underestimated in field experiments in shrublands? *International Journal of Wildland Fire* **28**(2), 127.

Poggi, D, Porporato, A, Ridolfi, L, Albertson, JD, Katul, GG (2004) The effect of vegetation density on canopy sub-layer turbulence. *Boundary-Layer Meteorology* **111**(3), 565–587.

Pope, SB (2000) *Turbulent Flows*. Cambridge: Cambridge University Press.

Raupach, MR (1994) Simplified expressions for vegetation roughness length and zero-plane displacement as functions of canopy height and area index. *Boundary-Layer Meteorology* **71**, 211–216.

Raupach, MR (1995) Corrigenda. *Boundary-Layer Meteorology* **76**(1–2), 303–304.

Raupach, MR, Finnigan, JJ, Brunei, Y (1996) Coherent eddies and turbulence in vegetation canopies: The mixing-layer analogy. *Boundary-Layer Meteorology* **78**, 351–382.

Rothermel, RC (1972) *A Mathematical Model for Predicting Fire Spread in Wildland Fuels*. Research Paper INT-115. Ogden, UT: U.S. Department of Agriculture, Intermountain Forest and Range Experiment Station.

Shaw, RH, Brunet, Y, Finnigan, JJ, Raupach, MR (1995) A wind tunnel study of air flow in waving wheat: Two-point velocity statistics. *Boundary-Layer Meteorology* **76**(4), 349–376.

Shaw, RH, Patton, EG (2003) Canopy element influences on resolved- and subgrid-scale energy within a large-eddy simulation. *Agricultural and Forest Meteorology* **115**(1–2), 5–17.

Shaw, RH, Schumann, U (1992) Large-eddy simulation of turbulent flow above and within a forest. *Boundary-Layer Meteorology* **61**(1–2), 47–64.

Smith, F, Carson, D, Oliver, H (1972) Mean wind-direction shear through a forest canopy. *Boundary-Layer Meteorology* **3**(2), 178–190.

Stull, RB (1988) *An Introduction to Boundary Layer Meteorology*. Fordrecht, The Netherlands: Kluwer Academic Publishers:

Su, H-B, Shaw, RH, Paw, KT, Moeng, C-H, Sullivan, PP (1998) Turbulent statistics of neutrally stratified flow within and above a sparse forest from large-eddy simulation and field observations. *Boundary-Layer Meteorology* **88**(3), 363–397.

Sullivan, AL, Knight, IK (2001) Estimating error in wind speed measurements for experimental fires. *Canadian Journal of Forest Research* **31**(3), 401–409.

Taylor, SW, Wotton, BM, Alexander, ME, Dalrymple, GN (2004) Variation in wind and crown fire behaviour in a northern jack pine black spruce forest. *Canadian Journal of Forest Research* **34**(8), 1561–1576.

Tieszen, SR (2001) On the fluid mechanics of fires. *Annual Review of Fluid Mechanics* **33**(1), 67–92.

Turner, MG, Romme, WH (1994) Landscape dynamics in crown fire ecosystems. *Landscape Ecology* **9**(1), 59–77.

Van der Hoven, I (1957) Power spectrum of horizontal wind speed in the frequency range from 0.0007 to 900 cycles per hour. *Journal of Meteorology* **14**(2), 160–164.

Van Wagner, CE (1977) Conditions for the start and spread of crown fire. *Canadian Journal of Forest Research* **7**(1), 23–34.

8

Coupled Fire–Atmosphere Model Evaluation and Challenges

JANICE COEN, MIGUEL CRUZ, DANIEL ROSALES-GIRON, AND KEVIN SPEER

8.1 Introduction

Wildland fires can have a positive effect in many ecosystems, by recycling nutrients, increasing diversity, and maintaining and regenerating critical habitat. They also endanger communities, especially as communities spread deeper into environments in which fires have been excluded. Variability in the climate system brings extrema such as droughts that allow wildland fires to penetrate ecosystems that are unprepared for fire, with devastating results. Predicting the spread of wildfires and bringing controlled or prescribed fire operations to the forefront of wildland fire management in challenging environmental conditions has required developing more sophisticated models and an understanding of when and why they work or do not work. Weather, fuels, fire, and topography all interact in a complex coupled dynamical system, and capturing the inherent spatial and temporal complexity of wildland fire in modeling tools is critical for managing fires in complex terrain and rapidly evolving weather and fuel conditions. Developing tools to manage fires, from the application of sophisticated ignition patterns during fire operations, fuel treatments and control lines, and efficient suppression will depend on access to models capable of handling high spatial complexity. Model simulations can also serve as an important tool to support communication of complex issues during training or to stakeholders who may or may not have a scientific background.

Coupled fire–atmospheric models are developed by explicitly linking either a physical or empirical model of fire processes in the atmospheric boundary layer with a computational fluid dynamics (CFD) model of the atmosphere (see Lopes et al. 1995; Clark et al. 1996a, 1996b; and Linn 1997 for some of the earliest implementations of coupled fire–atmosphere models). Coupled fire–atmosphere models have the potential to improve our understanding of key processes driving wildland fire dynamics and to enable better predictability of fire behavior across a wide range of surface and meteorological conditions. However, individual coupled models are unable to simultaneously resolve all the relevant scales, from topography and fuel structure, boundary layer physics, and up to larger scale environmental atmospheric conditions in coupled form, thus several types of models have emerged for applications at different scales. These may be broadly divided into models that put greater detail into fire processes, using grids of size approximately 1 meter but with less detail on atmospheric processes, and those that use larger grids of 100 meters to 10 kilometers scale, modeling processes that shape the atmospheric state in greater detail, but relying on parameterizations of fine-scale fire processes that are too small to be resolved. The challenges of evaluating each type are addressed here.

Modeling tools for predicting wildland fire behavior have increased dramatically in their ability to represent nonlinear feedback between the atmosphere and combustion environment (Hoffman et al. 2018). As coupled fire–atmospheric models are increasingly used in wildland fire science and management there have also been increasing calls for greater levels of model validation (Alexander and Cruz 2013a, 2013b; Cruz et al. 2017). However, there are significant challenges to model validation owing to the degree of data complexity required to initialize and evaluate coupled-atmospheric models (e.g. Linn et al. 2012) and interpreting landscape-scale fire environment and detection data (Coen et al. 2018a). The manner in which the fire environment and metrics of the fire itself have been recorded in the past are not adequate, notably in comprehensiveness, for the evaluation of current coupled models. Although there has been increased recognition of these challenges, a gap between the requirements for coupled fire–atmosphere model validation and the data or experiments used to conduct model evaluation persists.

Numerous research questions revolve around the continuous interplay between the wind in and near the combustion zone, the fire's thermo-kinetic activity contributing to rate of spread, and macroscale properties noted by observers – the fire line width, depth, shape (which may display shifting heading, flanking, and backing regions), and transient dynamic effects such as the formation of vertically-oriented vortices and their role in fire propagation. Coupled models have led in this regard, with little corroboration to date from measurements. An example of this situation is the dependence of spread rate on fire line width in an idealized small fire simulation, with uniform fuels on a flat surface, suggested to be due to lateral wind divergence reducing the head fire wind for a smaller width (Canfield et al. 2014). To evaluate this suggestion a high-resolution model must be confronted with high-resolution measurements, leading to the proper attribution of simulated fire behavior to wind divergence rather than some other aspect of the wind or fuels.

Other studies with coupled models explored fluid dynamical explanations for the emergence of the elliptical fire line shape (Clark et al. 1996a, 2004) and variations in shape depending on fire line length (Clark et al. 1996b) as short fire lines, for instance, a few hundred meters long, bow into a single elliptical shape under a single plume, while lines longer than about 1 kilometer split into several heads due to feedbacks between the hot plumes and near-surface wind convergence at the fire line. Vertically inclined vortices created by the fire may touch down and break up the fire (Clark et al. 1996a) or create rapidly spreading fingers of flame that shoot forward ahead of the fire line. Basic differences arise from a roll-dominated boundary layer versus convective boundary layer, and Sun et al. (2009) explored the interactions between boundary layer turbulence and fire spread. For the most part, these studies reference anecdotal or photographic evidence of these diverse aspects of fire behavior.

Many lessons regarding necessary components of a picture of the fire–atmosphere system consistent with observations have accrued over the history of fire science – for example, that key information, in fact, lay in the evolving shape of the fire front (e.g. Cheney et al. 1993), while fire–atmospheric interactions were recognized as an important part of mass fire behavior a half-century ago (Countryman 1969). Currently, high-resolution observations of wind and fuel within the three-dimensional combustion zone at the flame-scale are being more widely collected outside of the laboratory as existing meteorological instrumentation and analysis methods are applied to wildland fire field experiments (Charland and Clements 2013; Clements and Seto 2015; Liu et al. 2019). Other approaches to investigate this interplay estimated winds within flaming combustion zones, fire progression, and, in some cases, accompanying fire heat fluxes, from analysis of high-speed infrared (IR) imager data on the ground (Clark et al. 1999; Coen et al. 2004) or airborne (Radke et al. 2000; Stow et al. 2019; Schag et al. 2021) platforms. Though many of the latter observations were gathered from fires of opportunity in uncontrolled conditions, the adaptation of image flow analysis techniques to the analysis of IR observations allowed documentation of model-predicted phenomena such as the formation and role of vertically-inclined vortices in fire line dynamics (Clark et al. 1999) and the generation of forward-shooting fingers of flame (Radke et al. 2000; Coen et al. 2004). Finally, we want to understand how such physical effects scale up to exert significant control on larger fires. The answer is not yet clear.

This chapter is oriented toward model evaluation, the challenges that make this a daunting task, and the corresponding measurements needed to carry out an evaluation at some level. We use the word evaluation to mean the overarching effort to confront numerical simulations of coupled fire-atmosphere dynamics and effects with observations, which include the accompanying meteorological, combustion, and ecological conditions. Within that overall effort, a multitude of steps may be required for progress, involving validation, verification, calibration in appropriate and documented roles, and other tests. Much model validation occurs in the development and implementation of numerical schemes and parameterizations (e.g. Aumond et al. 2013). Does the model produce plausible results compared to either an established model in certain specific conditions, some or all expected airflow features or observations? When is disagreement with observations a sign of model failure versus inappropriate application or insufficient specification of inputs? Which level of accuracy is needed for a particular purpose? Delivering accurate forecasts or hindcasts of large, evolving, fires can require telescoping levels of comparisons and re-evaluations.

Evaluation efforts have been ongoing in the meteorological community for decades, with coordinated efforts to intercompare and evaluate models (Climate Modeling Intercomparison Projects or CMIPs, or, more closely related to the wildland fire–atmosphere coupled problem discussed here but at global scales, the FireMIP project, Rabin et al. 2017). Numerous measures of skill with spatial and temporal comparisons have been developed for difficult forecast problems such as precipitation (Gilleland et al. 2009). The nonlinear processes represented in coupled fire–atmosphere models imply that interactions between the separate parameterizations can give rise to unexpected behavior.

Part of model validation involves an assessment across all subgrid-scale parameterizations and their interactions. Prescribed fire experiments will help with the assessment of subgrid scale parameterizations but, as discussed in this chapter and elsewhere, the strongly chaotic nature of small scales poses further challenges. For instance, the recognition of stochastic behavior and coherent structures or dynamical modes in fire–atmosphere interactions has emerged from a more detailed understanding of the role of the turbulent atmospheric boundary layer in fire behavior (Simpson et al. 2016; Thurston et al. 2016) and, here, the detailed properties of the dynamic CFD must be considered (Coen 2018). Moreover, feedback on local meteorology from the fire can vary over a large range of phenomena, from deep convection and thunderstorm effects, enhanced winds, topographic channeling, and modified sea-breeze circulation (Peace et al. 2015; Coen et al. 2018a, 2018b). These complexities produce challenging problems in model evaluation.

Tools for model evaluation from diverse disciplines are being applied to fire simulations. Filippi et al. (2014) proposed a series of scoring methods to rate wildfire simulations, using mathematical techniques from biology, ecology, and image analysis, and applied them to research simulations of a large fire (Filippi et al. 2018). Assessment of forecast skill using object-based verification techniques (Davis et al. 2006) have advantages over point-to-point comparisons. Methods to measure agreement with observations are also done in real-time for better operational forecasts (de Gennaro et al. 2017; Cardil et al. 2019). Quantifying agreement with data is fundamental to data assimilation techniques, which then use such measures to update model parameters to improve the fit (Mandel et al. 2009; Zhang et al. 2017, 2019). A growing call for ensemble methods appears in many studies, as the modeling community grapples with uncertainties in both models and data.

We now recognize that, like models, data products also require scrutiny as they may be the result of algorithms or submodels, or themselves be a reanalysis product (for instance disparate data integrated by a model using a data assimilation system). Yet, until relatively recently, wall-to-wall initialization and validation data were not available for evaluation of simulations at landscape scales. For example, the advent of the Landscape Fire and Resource Management Planning Tools Project (LANDFIRE) database from the U.S. Department of Agriculture Forest Service and U.S. Department of Interior (www.landfire.gov; last accessed November 24, 2021), made spatial fuel model information sufficient for initializing many types of fire behavior models available across the United States by combining remote sensing and submodels for ecosystem simulation and biophysical gradients. Recent advances in satellite active fire detection at scales and revisit frequencies relevant to fire behavior such as the Visible and Infrared Radiometer Suite (VIIRS), accompanied by algorithmic development (Schroeder et al. 2014), now provide regular, wall-to-wall information on fire occurrence and extent, subject to known limitation of obscurance by clouds, view angle departure from nadir, and topographic blocking. Given these challenges, model developers, fire scientists, and managers who may use the results of these models are left wondering what methods and approaches can be used to critically assess and build confidence in their application.

With validated models and data one may carry out an evaluation of simulations, but, as alluded to above, the evaluation process of fires at smaller scales, in particular, is itself not well defined, due to the relatively more important contributions of small-scale turbulent structures as the resolution increases. Direct numerical simulation of wildfires down to molecular scales does not exist, thus all CFD models parameterize physical processes at some scale, including turbulent energy and momentum transfers. Provided that their numerics maintain sharp gradients and do not strongly damp small scales (Coen 2018), very high resolution (under 100-meter grid spacing) eddy-resolving coupled models of smaller fires may resolve some small-scale turbulent features or characteristics of the fire (e.g. Mell et al. 2007). Deterministic prediction of location and timing of individual surges in a fire line driven by turbulent boundary layer eddies is not predictable (Mukherjee et al. 2016), for example, but modeling the probability of such surges in order to implement a parameterization of their effects on fire spread may be a reasonable goal.

Developing parameterizations for such small-scale effects require data at correspondingly small scales. Initializing models at higher and higher resolution also requires higher resolution data. The detailed data on surface fuel properties and the atmospheric conditions required to initialize coupled fire–atmospheric models for small experimental fires are not typically available for historical data sets and can be extremely difficult and costly to quantify in new experimental campaigns (Linn et al. 2012). The high-resolution process models initialized with such data might be evaluated over short periods, but as the resolution increases and the heterogeneity of the fuel emerges, the representation of turbulence and combustion in the model also comes to the forefront of the problem and may determine which parameters can be usefully evaluated.

Central to the notion of model evaluation are the objectives of the modeling or measurement program. Operational models or simulations will have different objectives than will research or process models and simulations, with detailed fire spread metrics and smoke movement of critical importance to the former due to their operational importance while capturing overall event evolution and appropriate sensitivity to parameters can sometimes be more important to the latter. Process models may be less focused on rates of spread as the end goal and be evaluated over a narrower range of parameters where targeted effects can be resolved and new phenomena investigated. It is important to keep objectives in mind when evaluation tools are being used or interpreted for different communities.

This chapter presents an evolutionary view of validation philosophy and methodology. We start in Section 8.2 with more traditional methods that compare fire front propagation to models, whether or not those models directly couple atmospheric processes to the spread rate; empirical methods, for instance, may have some coupling effects built in to the spread rate through observational (laboratory or field) constraints. Next, we examine fuel considerations on model error and the uncertainty of coupled fire–atmosphere simulations due to fuel representation (Section 8.3). This is followed by the new emerging paradigm of the effects of turbulence and coherent flow structures in fire dynamics, from the impact of atmospheric turbulence and small scale processes to intrinsically coupled effects, represented in landscape coupled models in both hindcast and forecast modes (Section 8.4). Finally, Section 8.5 summarizes and provides a discussion of benchmarks, metrics, and current research issues that we believe should be addressed for model evaluation to advance.

8.2 Fire Front Spread Rate Expectations and Uncertainties from Models and Data

The evaluation of complex models comprises several components that depend on the model framework and specific application, but could include components such as conceptual validity, sensitivity analysis, comparison with other models, extreme condition tests, and direct comparison with observed (e.g. in experimental fires or wildfires) or modeled (by existent accepted models) variables (Rykiel 1996; Jakeman et al. 2006; Alexandrov et al. 2011). All these aspects should be considered in a model evaluation framework. Discussing the full gamut of steps within a comprehensive evaluation (e.g. Blocken and Gualtieri 2012) is beyond the scope of this chapter, and we will focus on the direct comparison of model predictions with observations, and a discussion of challenges that fire modeling faces as models confront data at many different levels. We seek to emphasize the quantitative expectations from models and the uncertainties present both in model physics and data. In this section the discussion centers on the traditional notion of the rate of spread of a fire front.

A key feature of a wildfire is that, in moderate to low winds and uniform fuels, with flat terrain, it initially spreads on the landscape as a simple parabolic front, called the head. More often the front is not a simple curve but made up of surges in multiple fire fingers (e.g. Clark et al. 1996a). In very high winds or for some fuels there may not even be a well-defined front, as embers ignite broad areas of fuel ahead of the main front, a process known as spotting. In strong winds, the rapid spread will lead to large burned areas and possible widespread destruction of human life and property (Figure 8.1; Cruz et al. 2012; Coen et al. 2018a). In these circumstances, not surprisingly, the forward spread rate of a fire, whether easy or not to define, has been a key variable being studied by researchers (Fons 1946; McArthur 1967; Albini 1976; Mell et al. 2007; Linn et al. 2012). And in light of recent catastrophic events, with an increase in the number of large fires over recent decades (Dennison et al. 2014), and human fatalities (Cruz et al. 2012; Lagouvardos et al. 2019), the ability to predict the spread rate of a wildfire and its first time of arrival to specific locations can be seen as a crucial need for agencies responsible to manage fires and keep the public informed of potential threats. While one can also argue that the evaluation of a Coupled Fire–Atmosphere Model (CFAM)'s predictions of flame front movement be a fundamental test for operational relevance, at landscape scales, additional factors are significant (discussed in Section 8.4), such as evolution of the fire's shape and production of fire and weather phenomena. At the same time, as noted in Section 8.1, for many applications in prescribed fire or in ecological studies of fire effects, other aspects of modeling efforts are as important, for instance fire intensity, smoke production, degree of vegetation consumption, and so on.

The forward rate of spread appears to be one of the easiest fire quantities to measure, requiring in its simplest form the time of arrival of the flame front at two distinct locations. Even for relatively uniform conditions, however, the rate of fire spread in the field is not constant, varying widely in time and space, as the fire responds to changes in wind speed and direction, fuel condition, and topography

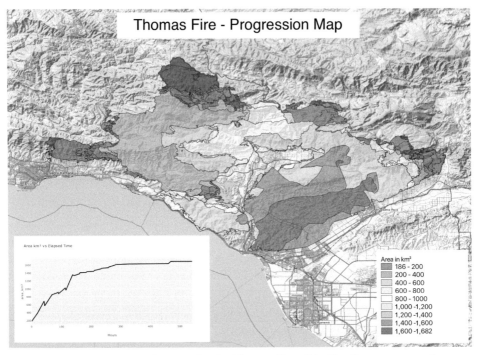

Figure 8.1 Fire progression map for the Thomas Fire, CA. Shading represents time over the period December 5–24, 2017. Note the complex evolution of the fire front boundary and the spread pattern, as it realigns with the changing wind and local topography. According to nearby RAWS the wind initially was gusting about 30 mph to the NE, with RH about 8%. Inset shows cumulative area (square kilometers) burned as a time series in hours from ignition. Data from GEOMACS A black and white version of this figure will appear in some formats. For the color version, please refer to the plate section.

(Figure 8.2; see also Clark et al. 1996b for similar variations in spread rate in a CFAM). In addition, fires do not always spread perpendicular to what appears to be the fire front; thus, connecting before and after locations for a point along the line can be challenging. Experimental fires conducted with the aim of measuring fire behavior and the environmental variables driving fire propagation have been a primary source of data for conducting simple evaluation of small-scale CAFMs. Cruz and Alexander (2013) detail 48 published datasets used to evaluate empirical fire spread models to characterize the error metrics expected to occur when using this type of model. Individual fires from some of these, namely Cheney et al. (1993) and Stocks et al. (2004), have been used to compare CAFMs predictions of rate of fire spread with observed ones (e.g., Mell et al. 2007; Linn et al. 2012). The advantage of these experimental fires is that their small size, compared to a typical wildfire, and relatively homogeneous environment (often no slope and

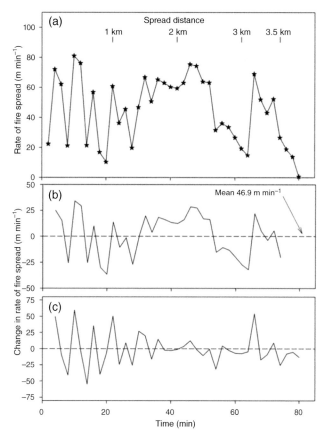

Figure 8.2 Rate of spread and variability in head fire rate of spread at 2-minute intervals versus elapsed time since ignition for an experimental fire in grassland (spinifex) fuels in Western Australia (adapted from data presented in Burrows et al. 1991): (a) rate of fire spread as a function of time from ignition to extinction; (b) changes in the rate of fire spread about the mean (46.9 m min^{-1}); and (c) change in rate of spread from the previous 2-minute time interval. From Cruz and Alexander (2013)

fairly uniform fuel properties) constitute a simple case to start exercising CAFMs under conditions of lower uncertainty. The disadvantage of relatively small experimental fires for CAFMs evaluation is that detailed atmospheric motions at small scales are mathematically unpredictable after a short time, that is, the error growth in turbulent CFD models increases exponentially with time (Lorenz 1969). In practical terms, a deterministic simulation of their growth is unable to accurately simulate if, when, and where an eddy will advance or oppose a particular segment of the fire line, driving comparisons toward a statistical framework.

The evaluation of CAFM that describes a large number of physical processes and their interrelations is obviously more complex than the evaluation of simple one-dimensional (distance versus time) fire spread models (Cruz and Alexander 2013). CAFM simulations include a spatial component that describes the evolution of the fire perimeter with time and a three-dimensional component that describes the interaction between the fire energy released and the ambient environment. As such, the movement of the flame front is only one of the results of a CAFM simulation, with other outputs such as energy released, fluid flow velocities within and above the flame, heat transfer into unburned fuels, defining the fire itself and determining the overall spread rate of a fire. One of the common advantages pointing to the use of CAFMs is their ability to incorporate the feedback between the fire and the atmosphere, allowing us to better capture the flow field driving fire propagation.

As has become clear – especially from coupled model studies – rate of spread is not a simple single number or even a well-defined frontal motion, but a stochastic process resulting from multiple feedbacks on wind and fire intensity (Simpson et al. 2016). New metrics are needed to define and explore fire spread in more complex environmental conditions. To illustrate the idea of exploring spread metrics and the numerical sensitivity to basic wildland parameters, we have constructed a set of runs using the intermediate CFAM model QUIC-Fire. This model couples a cellular automata form of fire spread to a simplified atmospheric model (Linn et al. 2020). Runs with variable wind (specified uniform upwind value), fuel load, and fuel moisture were made on a domain of 400 × 400 meters. Typical experiments ran for a total of 300–400 seconds. Only grass-type fuel conditions were considered here, without any canopy or shrub vertical structure.

Examples of large-scale metrics that describe distinctive characteristics of the fire perimeter include reduced area (burned area divided by burned perimeter squared), which measures the fingering of the fire, and rate of spread divided by bulk rate of spread (RoS/BRoS), which indicates the amount of forward spread compared to spread in all directions. These metrics give a numerical value, for example, to behavior associated with point-shaped and parabolic-shaped fire (Cheney et al. 1993). The RoS/BRoS can show transitions and changes in intensity, runs, and other effects linked to evolving environmental conditions (Figure 8.3).

Data from field experiments designed to produce testable observations are the next step toward realistic model evaluation. In the following we examine some of the available datasets that have been used to test models, and continue in Section 8.4 to discuss issues and comparisons of model results to actual wildfires in several case studies. As with model sensitivities, the challenge of comparing models to data from the field involves defining suitable metrics, simple enough to be robust yet providing useful detailed information about model performance. Spread rate typically forms an initial stage of the process of model evaluation.

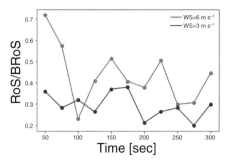

Figure 8.3 Rate of spread divided by bulk rate of spread (RoS/BRoS) for two CFAM (QUIC-Fire) experiments with different wind speeds. Periods of slow RoS transitioning in less than a minute to faster headfire runs are apparent, more so in the case with higher wind speed. A key model parameter determining the time scale of fire–wind interactions is the fuel burnout time from ignition, set here to be 30 seconds.

The quality and detail of the existing datasets varies, and it is important to have a pragmatic approach when applying CAFMs to experimental fire data. There will be CAFM input that might not be described at the scales or resolutions optimum for the model, and best assumptions will need to be made, as discussed by Linn and Cunningham (2005), Mell et al. (2007), and Marino et al. (2012). From the empirical evidence that wind speed and dead fuel moisture can explain up to 70% of the variation in the rate of fire spread in an experimental dataset with homogeneous fuels and equivalent initial conditions, e.g. fixed ignition lines, (e.g., Stocks 1987, Cheney et al. 1993; Alexander and Cruz 2006), it might appear to be reasonable to expect similar controls for CAFMs in similar homogeneous conditions, if the models are capturing the key processes driving fire propagation. This assumes that the coupled processes are similar in the experiments, which may not always be the case. While such simulations cannot be expected to deterministically reproduce an individual fire's progression because of the mathematical limits to predictability to small atmospheric motions, they may instruct as to whether simulations capture the observed sensitivity in response to changes in input variables. Counterintuitively, model response and uncertainties may be better understood against more complex data, such as data emanating from wildfire case studies where strong topographic, wind, or fuel environmental forces drive fire propagation individually or in combination at various times and locations of the fire (e.g. Coen and Riggan 2014).

A description of selected datasets that are most commonly used when comparing models to data shows the diversity of scales, topography, and fuels involved. Table 8.1 provides a few examples of datasets that could potentially be used as benchmarks to exercise CAFMs under a broad range of vegetation

Table 8.1 *List of experimental burning projects with data for use in evaluation of CFAMs*

Vegetation type (# of experiments)	Typical plot size	ROS range (m s^{-1})	Source	Observations
Grass (121)	Variable, mostly 100 m × 100 m, but up to 200 × 200 m	0.29–2.07	Cheney et al. (1993); Cheney and Gould (1995)	Fuel structure controlled for height and load; Wind speed and direction measured at 2-meter height with synchronized anemometers at the corners of the experimental plots. Ignition pattern documented.
Grass (26)	33 m × 33 m	0.05–1.70	Cruz et al. (2015)	Paired experiments with control of degree of curing; Detailed information on fuel, percentage of dead fuel, moisture content of live and dead fuel. Experiments cover a broad range of overall fuel moisture contents. Wind speed and direction measured at 2-meter height at two locations upwind from the ignition line. 3D sonic anemometers used to characterize turbulence.
Grass (1)	450 m × 600 m	1.2	Clements et al. (2007); Clements (2010); Dupuy et al. (2014)	Wind measurements conducted at various heights on 43-meter and 10-meter tall towers located inside the burn unit.
Shrubland	100 m × 200 m	0.07–0.92	Cruz et al. (2013)	Semi-arid vegetation with horizontal and vertical discontinuities. Wind speeds measured at 10-meter and 5-meter heights in the vicinity of the experimental block. Burns conducted over a period of 3-years across a range of fuel dryness.
Boreal forest (11)	mostly 150 m × 150 m	0.26–1.16	Alexander et al. (2004); Stocks et al. (2004); Taylor et al. (2004)	High intensity crown fires conducted in jack pine stands with a black spruce understory. Wind measurements conducted in a fixed location within the

Table 8.1 (*cont.*)

Vegetation type (# of experiments)	Typical plot size	ROS range (m s^{-1})	Source	Observations
				experimental area. Fire behavior data includes vertical profiles of flame temperatures and radiative heat fluxes. Fluid flow estimated from IR imagery (Clark et al. 1999).
Eucalypt forest (99)	200 m × 200 m	0.01–0.38	Gould et al. (2008); McCaw et al. (2012)	High intensity fires in eucalypt forests across a broad range of fuel structures. Burns conducted under dry summer conditions. In-stand wind dynamics measured at four anemometers located upwind from experimental plots. Data includes detailed vertical flame temperature profiles (Wotton et al. 2012).
Forest and grassland (16)	Small to large (10–1,000 ha)	0.23–0.44	Butler et al. (2016); Clements et al (2016); O'Brien et al. (2016); Ottmar et al. (2016)	Highly instrumented operational prescribed burns. Detailed analysis of fire, atmosphere, and smoke dynamics described in several scientific papers (see Ottmar et al. 2016).

structures and burning conditions. However, even large experiments that have required multi-agency cooperation have been unable to capture the quantity and quality of information required to initialize and evaluate more than a few aspects of physics-based models.

As these datasets demonstrate, it has been historically assumed by the fire community that, for purposes of simulating and experimental burn, the atmospheric state – arguably along with fuels one of the most important factors affecting fire behavior – may be adequately specified by several surface measurements of wind speed, limited tower measurements, and occasional nearby lower atmosphere vertical profiles. This assumption aligns with the simple input needs of historical, less complex models, and a single profile may suffice as the minimum atmospheric information required to start off a CAFM configured for a small-scale numerical experiment. However, unless conditions are very homogeneous, flat, laminar, and unchanging for the duration of the fire and thus not subject to incoming or downward moving atmospheric boundary layer effects, these datasets omit aspects of the atmospheric state that will fundamentally control the fire. An alternative method for initializing the atmospheric state of a CAFM uses a four-dimensional gridded atmospheric analysis to initialize the modeling volume, provide boundary conditions, and perhaps to nudge interior points toward observed values. This approach can introduce large-scale meteorological events and gradients that shape the weather around the fire into the modeling domain. On the other hand this approach too is not optimal, as the vertical and horizontal resolution of the analyses are relatively coarse, so prescribed fires often fit, at least initially, within a single grid volume or line, and atmospheric vertical structure – a key factor for plume rise – is not well resolved.

There are inherent uncertainties in the datasets in Table 8.1. With few exceptions, the datasets were not developed to support the development, calibration, or evaluation of CAFMs. The CAFM community together with the observational community have made major strides but are still some distance from the needed combined effort to develop a systematic assessment of the uncertainties in data collection, or to describe the data requirements for CAFM usage. The application of CAFMs to these datasets will require the models to adjust to the sampling characteristics of the experiment (e.g. temporal and spatial scales of the experiment, range in intensities) and the uncertainty, or variability, associated with a number of the measured variables (e.g. the definition of fuel characteristics, distribution, and gradients in fuel moisture content, wind spatial and temporal structure). Methods will need to be developed to overcome certain data characteristics (e.g. wind measured at fixed x, y, z locations upwind from the fire) or a dominance of descriptive rather than quantitative components of datasets.

As the model treatment of fuels is a simplification of its real characteristics, quantitative synthetic measures are needed to describe the existing fuel arrangement. As an example, a fuel complex dominated by five different plant species with distinct geometric structures and diverse fuel moisture content needs to be simplified for models whose resolution is at a scale larger than the variation within the fuel bed. The need to represent fuel descriptors such as surface area-to-volume ratio, porosity, and wind drag, at the model subgrid-scale is a well-known model problem. The conversion of microscale observations to parameters useful for coarser scale models requires, as these measurements become more common, a specialized form of data analysis and evaluation. For example, do mean values adequately represent a fuel complex? Does a fire propagating through a grass field measurably spread differently if heavy larger fuel elements are scattered across it? Errors are often assumed to have a normal distribution, yet fire physics abounds with threshold behaviors, for example a location is either on fire or it is not. Such fuel-related issues are discussed further in Section 8.3.

To overcome some of these difficulties and uncertainties that can impact the model performance, modelers can conduct direct comparisons of CFAMs outputs with those from traditional models, namely empirical-based or semi-empirical fire spread models (e.g. Rothermel 1972; Cheney et al. 2012). Understanding a highly simplified component of CFAM behavior, namely whether or not the model accurately represents some recognized empirical relationship between an environment variable (e.g. wind speed, fuel moisture, fuel bed depth) and model outcomes, or the sensitivity of the model to input variables, can be gained from such analysis (e.g. Marino et al. 2012; Moinuddin et al. 2018). In fact, this kind of comparison is commonly made for models of intermediate complexity (e.g. Filippi et al. 2014).

To help untangle sources of uncertainty in CFAMs, it can be helpful to carry out sensitivity experiments. Unfortunately, this information can be difficult to synthesize and apply to model simulations without distracting from the intended application of the model. A simple example of numerical sensitivity studies was performed with QUIC-Fire by varying several basic parameters individually and then together, to get a maximum co-varying response. Figure 8.4 shows that the total burned area in a given time can easily vary by a factor of two for modest changes in the three primary control parameters (wind speed, fuel moisture, fuel density). Note that we distinguish sensitivity experiments from ensemble approaches. Sensitivity experiments may test the entire natural range of possible inputs, whereas ensembles are intended to tell something about the probability of a particular outcome. Not all possible inputs are equally likely.

A different approach, again motivated by the need to simplify the problem to a manageable level, is to use a statistical method to fit a fireline to observations, such that both the simulated fire spread rate and atmospheric conditions are modified to

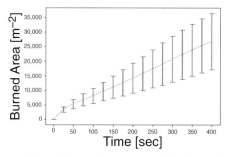

Figure 8.4 Sensitivity of total burned area as a function of time for a CFAM (QUIC-Fire) to variability in three primary parameters: fuel moisture content (3–7%), fuel density (0.5–0.9 kg m^{-2}), and wind speed (4–8 m s^{-1}). Error bar displays 90% of the variation

drive the fire front to agree with observations (e.g. Zhang et al. 2019). Although not coupled in the usual sense these models essentially represent evolving feedback between the fire and the environment statistically, typically using wind, though other parameters such as fuel characteristics, humidity, solar heating, etc., are possible too.

While local metrics and point-to-point model-data comparison to observation are crucial for the energetics and extreme behavior, such comparisons are fraught with uncertainty. The burning fire front (observed by remote sensing, either airborne or satellite) approximated as a one-dimensional curve is the standard dataset used for initialization and model evaluation. In the presence of stochastic effects due to underlying dynamical instabilities, the methodology of evaluation moves toward statistical measures of agreement. These statistical measures can be local or integral. Spread distance at points on the perimeter (Fujioka 2002) or burned area at given times (a burn area "snapshot") is typically used for spatial comparisons to assess model performance (Coen et al. 2018a; Filippi et al. 2018), and can reveal agreement or not during growth periods.

Ultimately the statistical approach has led, in the atmospheric community, to more sophisticated methods of data assimilation and state estimation, that are capable of representing the physics of the system and at the same time agreeing with observations within some limits. But it is also necessary to develop new metrics of integrated structural and behavioral aspects of the fire.

If the evaluation of a CAFM is more than the evaluation of the predictions of flame front movement, the challenge is to define what processes and outputs should be evaluated. Following a bottom-up approach, the first question pertains to the smallest scale at which the model should be verified. For models with fine-scale chemistry and thermodynamics the pyrolysis, combustion, and heat release components of the model should be evaluated (Sullivan 2017; see also Chapter 1). Similarly, model outputs such as fluid velocities and temperatures within and around the flaming front, and heat transfers into unburned fuels need to be evaluated, as they have been in laboratory conditions (Morandini et al., 2005; Balbi et al., 2010). The representation of the initial lofting and propagation of embers could very well depend on such fine scale dynamics and turbulence (see also Chapter 5).

The intrusive nature of heat transfer sensors with bulky volumes and flat surfaces (e.g. Frankman et al. 2013) implies that they do not necessarily reproduce exact fuel particle boundary conditions, but rather a filtered version of fluxes. Such considerations limit in principle the evaluation of a model heat transfer formulation. Measurements of heat transfer quantities and flame fundamental properties have shown a large spatial variability associated with the highly turbulent and chaotic environment of a fire (e.g. Butler et al. 2004, 2016). Quantitative measures that address this variability, as well as the number of

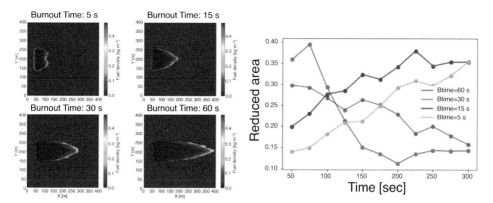

Figure 8.5 (Left) Fuel density for a homogenous grassland fire in a CFAM (QUIC-Fire) displayed at 300 seconds for four different burnout times, representing four fine fuel categories (30 seconds is a typical burnout time, or e-folding time scale for complete combustion of a model fuel cell element). (Right) reduced area as a function of time for the four burnout times; the two longer times show a decrease over time, while the two faster times show increasing reduced area. Reduced area decreases due to enhanced fingering with slower burnout times.

A black and white version of this figure will appear in some formats. For the color version, please refer to the plate section.

measurements or sampling required to adequately describe the leading edge of the combustion interface are needed (see also Chapters 3 and 5).

Combustion processes are crucial to the air flow and fire progression within the fuel bed yet these processes are typically crudely parameterized even in fine-scale CFAMs such as FIRETEC and WFDS. Often some version of Arrhenius' law is used to introduce temperature dependence to the ignition process. In the simplified CFAM model used here for illustration, QUIC-Fire, the combustion is represented by a temperature-independent burnout time. A burnout time may represent different kinds of fuels as well as the rate of pyrolysis. Figure 8.5 shows that the reduced area varies dramatically depending on the model's specified burnout time, a simple representation of combustion on a grid. In this CFAM the burnout time is constant for the entire domain, though other fuel properties can vary. These basic responses of fire behavior to variations in bulk fuel parameters have been reproduced by less detailed NWP-based CAFMs (e.g. Coen et al. 2013). One of the strengths of tools such as FIRETEC and WFDS lies in an additional ability to test the effect of more subtle fuel variability such as due to beetle kill or vertical structure.

8.3 Fuel Considerations in Models and Validation

The strong dependence on fuel parameters in small or weakly-driven fires leads us to the problem of fuel representation in models, and the degree of fidelity necessary to produce realistic simulations. Fine-scale fuel structure is unlikely to be resolved at landscape scales, hence the need also to understand the key structural properties of fuel at various scales for potential parameterizations.

During the 1930s, fuels were described in basic concepts of time and difficulty to suppress (Hornby 1936). Rate of spread (as discussed in Section 8.2) was considered the most important metric of fire behavior since it is the most relevant to firefighter safety and was usually estimated by statistical analysis of individual fire reports. Although landscape level models such as CAWFE (Coen 2005, 2013), WRF-FIRE (Coen et al. 2013) and derivatives such as WRF-SFIRE (Mandel et al. 2011), and ForeFire/MesoNH (Filippi et al. 2009) continue to use the same level of fuel details as FARSITE (Finney 1998), that is, stylized fuel models, finer-scale CFD-based models such as WFDS (Mell et al. 2007) and FIRETEC (Linn et al. 2002) that parameterize combustion at finer scales are able to take advantage of more sophisticated fuel characteristics. The scale at which fuels are resolved in the models are highly dependent on the intended purpose and resolution of the model, with landscape models having fuel cell resolutions on the order of 30–100 meters, while some CFD fire models resolve fuels down to approximately a 1-meter scale. High spatial variability has been observed even in simple fuel beds (Hiers et al. 2009; Keane et al. 2012) accentuating the need to describe the actual fuel arrangement and composition as opposed to abstract representations.

Fuel loading and characteristics have long been considered a primary driver in fire behavior, with differences in species, composition, spatial structure, and moisture all interacting to determine probability of ignition, consumption, fire radiative power, and rate of spread. Fuel bed arrangements are complex and highly variable, characteristics and physical attributes that lead to a high variation in fire behavior and effects (Fahnestock and Key 1971; Hiers et al. 2009; Loudermilk et al. 2012). Fire scientists and modelers, along with fire managers, have historically led the charge to characterize and classify wildland fuel bed characteristics, but these variables and measurements, especially at finer scales of resolution, are becoming increasingly important in related fields such as fire ecology, air quality management, and carbon balance modeling (Sandberg et al. 2001). For nearly a century, fire control planning and firefighter safety have been the underlying drivers of fuels classification in the United States, which has led to a fuel representation that fits the dominant fire spread and fire behavior models, in particular those based on Rothermel (1972) and refined by Sandberg et al. (2007). As such, the characterization of fuels has been, to some extent, held back by the combination of the difficulty to accurately measure 3D vegetation structure, the limitations of model applications for which the fuels were represented, as well as the prohibitive cost of fuel characterization uniformly and repeatedly on a landscape scale.

To simplify fuel inputs, the Albini (1976) classification system as restated by Anderson (1982) specifies 13 stylized fuel models that would supply numerical fuel bed descriptions as simple and easily understandable inputs for the Rothermel fire spread model and result in fire behavior predictions that were reasonably accurate when compared to catalogued rate of spread. These standard models were meant to be easy to visualize and implement for the fire community. Fire behavior fuel models (FBFMs), such as Anderson's (1982), have been created for a variety of conditions, expanded (Fernandes et al. 2000; Scott and Burgan 2005; Bacciu et al. 2009; Santoni et al. 2011), and are still widely used. While they lack certain key measures of forests and fuels, such as floor depth and or woody fuel mass, they can be easily created and be fitted into a wide array of fire behavior results. On the other hand, they are hard to use in developing scenarios, and some researchers consider it impossible to assign or develop fuel models (Burgan 1987) without prior knowledge of fire behavior and spread in fuel types (Keane 2013). The Fuel Characteristic Classification System (FCCS) (Ottmar et al. 2007) expands fuel characterization and classification complexity by reporting fuels in six vertical layers (canopy, shrubs, nonwoody, woody, litter/lichen/moss, and ground fuels (i.e. duff) and fuel categories within each stratum). This level of abstraction in classification heavily relies on previous experience and expert opinions from local fire users in all phases of the fuel mapping process (Keane and Reeves 2012).

During the early 2000s, there was an emergence of a numerical modeling simulation of wildland fire behavior that incorporated computational fluid dynamics and strived to represent critical physical mechanisms of energy transfer and fire–atmospheric feedback responsible for fire spread (Mell et al. 1996, 2007; Linn 1997; Morvan and Dupuy 2001; Linn et al. 2002, 2005; Dupuy and Morvan 2005; Zhou et al. 2007). These models were designed to run on a much finer resolution, and required detailed 3D inputs for fuels, winds, and topographical information. In the early 2000s this demand was met with new approaches to fuel descriptions and measurements (Sandberg et al. 2001). Several methods were introduced that would strive to represent the vertical spatial variation of the fuels, with an emphasis of a realistic representation of the canopy, but limited attention was given to either the horizontal variation of surface or canopy fuels, or surface fuel vertical variation. These methods did not encompass the full range of heterogeneity and spatial variety within landscapes (Hardy et al. 2008), despite providing better estimates of vertical continuity. The Fuel Characteristic Classification System (FCCS; Ottmar et al. 2007; Riccardi et al. 2007) described vertical heterogeneity and arrangements across surface, midstory, and canopy in highly detailed fashion but assumed a constant horizontal continuity within all the fuel strata.

Fuel properties, such as loading (Brown and See 1981), heat content (Van Wagtendonk et al. 1998), size (Van Wagtendonk et al. 1996), and moisture (Agee et al. 2002), are all highly variable in both time and space. Fuel beds are incredibly complex and generally consist of multiple fuel types and fuel components (litter, grass, shrubs, logs, trees, etc.), with each fuel component having specific properties (loading, moisture, surface area-to-volume ratio, etc.). These specific properties vary at different spatial scales (Habeeb et al. 2005) For example, trees aggregate into patches at different length scales than grass. Furthermore, the variability that can be found within a relatively small stand can be as high as the variability across different landscapes (Brown and Bevins 1986; Keane 2008). Fuel loadings are so highly variable that they often have poor correlations to vegetation characteristics, topography, and climate patterns (Brown and See 1981; Cary et al. 2006).

Kalabokidis and Omi (1992) investigated fuel depth and loading within forest and sagebrush environments and found significant variance at a 60 meter scale. Reich et al. (2004) modeled fuel properties derived from remote sensing at a scale of 30 meters and found that spatial variability could be reasonably linked to topographic data and vegetation type. Fire adapted forests such as Ponderosa pine (Pinus ponderosa) are associated with fine scale heterogeneity (Stephens et al. 2015), but alternative fire regimes have often been found to be dominated by large-scale patches (Turner and Romme 1994; Johnson et al. 1998). Hiers et al. (2009) used LIDAR to measure small-scale variation in fuel heights in longleaf-pine dominated forests in the southeastern US and found that variation within the fuel-bed could be characterized by small-scale (0.5 m^2) patches, termed wildland fuel cells, which became spatially independent at larger scales, and found fire behavior varying at similar scales. These cells were also found to have unique nonlinear combustion characteristics (Loudermilk et al. 2012). Keane et al. (2012) used spatial variograms to quantify spatial variability of fuel component loading in six forest and rangeland ecosystems. Fuel components each had unique scales of variation, with fine fuels varying at horizontal scales of 1–5 meters, shrubs and midstory varying at 10–150 meters, and canopy fuels at scales of 100–500 meters.

At local scales, fire spread is greatly influenced by the three-dimensional structure and distribution of the fuel bed (Pimont et al. 2009; Rocca 2009; Parsons et al. 2011); an example of this is the control that small, fuel-depleted patches can exert on the direction and speed of spread of an incoming fire (Agee et al. 2000; King et al. 2008), or how dry ladder fuels can complicate a simple surface fire with the possibility of crowning. Regardless, most fuel mapping efforts assign large scale averaged fuel values to large areas (Keane et al. 2001; Reeves et al. 2009), assuming homogeneity at scales not seen in the natural world and ignoring the extent to which fuel variability can influence wildland fire behavior and effects.

Linn et al. (2002) explored thinning canopy fuels and found that the reduced drag within the canopy led to faster wind speeds at the surface. Fuel patterns were also explored regarding tree distribution, with domains with high-density zones leading to higher rates of spread than those with a more uniformly random distribution. Pimont et al. (2009) found that fire behavior was highly dependent on the structure and level of representation of the fuels in the model, with a 50% increase in rate of spread and a 200% increase in crowning potential when the fuel was represented as a vertically heterogeneous fuel as opposed to a single layer of fuel. Hoffman et al. (2013) explored the effect of bark beetle and tree mortality on fire behavior and found that spatial patterns showed a significant effect on rate of spread and crowning potential. Pimont et al. (2011), found contradicting results with tree clumping having little effect on rate of spread and fire intensity, even with an increase in wind field variability. Ziegler et al. (2017) examined seven different real-world fuel treatments and found that fire intensity and rate of spread actually decreased within the treatment zones, independent of wind speed. Parsons et al. (2018) used STANDFIRE to look at the effect of thinning in fire behavior models and found more intense fire behavior with greater consumption and rate of spread but lower tree mortality rates. Contradictory results lend credence to the notion that these scales and dynamics are highly related to local conditions of wind–fire coupling and small fuel bed scales.

We use QUIC-Fire again to illustrate the role of fuel conditions in an idealized coupled model. In low moisture conditions, the mean rate of growth of the burned area is not significantly different for different resolution levels, but the variance is significantly higher. For all moisture and wind conditions, very small resolution scales (few meters) led to much lower variance. In high wind conditions, all moisture conditions have approximately the same characteristics. Coupling the large difference in burned area for high wind conditions with a relatively similar maximum rate of growth indicates that lower levels of resolution (larger resolved scales) have much flashier fires, with large amounts of growth occurring in bursts of intense fire behavior, while high levels of resolution (smaller resolved scales) tend to have much more stable predictable fire behavior. This is a surprising behavior since lower levels of resolutions in the example here imply essentially homogeneous grass, where we do not expect high levels of variance, and suggests that wind–fire coupled effects may be acting more effectively. An example of the minimum, median, and maximum burn density scenarios for the high wind, relatively low resolution, low moisture condition burns is shown in Figure 8.6.

The representation of vertical structure of fuel beds with increasingly complex representation of trees and canopies has helped to improve fine-scale model realism, but the horizontal heterogeneity of surface fuels remains to be thoroughly explored.

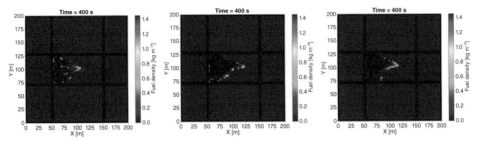

Figure 8.6 Fuel consumption at simulation end time (400 seconds). Fuel density remaining at the end of the simulation for the minimum (left), median (middle), and maximum (right) burned density for a low resolution, high wind, low moisture run. A black and white version of this figure will appear in some formats. For the color version, please refer to the plate section.

The scales at which fine fuel variability is relevant to fire behavior and fire effects is likely also to be highly dependent on the manner in which combustion is represented, the fire intensity, itself a result of coupled, nonlinear processes, and the local, evolving flow conditions. For landscape-scale CAFMs, discussed in Section 8.4, combustion in the boundary layer produces a spatially and temporally varying distribution of sensible and latent (mainly water vapor) heat fluxes, smoke, and lofted embers. Most aspects of the combustion process are parameterized in these models because combustion occurs at scales much finer than the spatial resolution of atmospheric models. More detail may not be necessary for answering landscape-scale research questions and applications, to which such models are best suited, as long as the spatial and temporal production of heat and threshold transitions (e.g. from surface to crown fire) are well represented. Developing such parameterizations is an important ongoing activity in fire modeling.

8.4 Hindcasting Case Studies and Challenges

Large wildland fires can occur over periods of a day to months, the largest spanning hundreds of thousands of acres (Andela et al. 2019), and within which are periods and areas of growth driven by differing fuels, moisture conditions, terrain, and microscale air circulation. CAFMs best suited to these events employ a Numerical Weather Prediction (NWP) model as the CFD component in order to represent weather systems, complex topographic effects on air thermodynamics and dynamics, radiation, cloud and surface processes and effects, along with a package of algorithms representing fire behavior. First, we ask what properties can be used to evaluate how good a landscape-scale wildland fire simulation is, separate from the observations that might be used to document them.

Three distinct requirements may be used to rate fire simulation outcomes. These are:

1. Objective assessment of expansion of the fire perimeter. This might, as described earlier in this chapter, be expressed simply by the rate of progression of the flaming front, or in terms of the rate of increase in fire-affected area between observation times. In light of their multi-day, large area, and episodic nature, simulations can be evaluated in terms of which area of the fire is advancing over given time periods. Typically, information on fire extent is derived from airborne or satellite infrared data or incident intelligence.

2. A combination of subjective and objective evaluation of the evolving shape, including where a fire turns, splits into multiple heading regions, accelerates, produces flank runs, or displays distinguishing features or variation along the shape. For example, multiple plumes occurred along the King Fire head region and bulbs of growth along its side (Coen et al. 2018b); two changes in direction occurred during the June 30, 2013 growth of the Yarnell Hill Fire (Coen and Schroeder 2017) (Figure 8.7); during the Esperanza Fire, the fire split into two heading regions – a plume-driven heading region drawing itself uphill and a wind-driven heading region running across the lower slopes (Coen and Riggan 2014); and during the Tubbs Fire, despite northeasterly winds, the southeast flank bulged as it drew itself up a hill (Coen et al. 2018a) (Figure 8.8). Unlike other bulk methods to evaluate instantaneous simulated atmospheric wind, this more dynamic approach reflects a cumulative tally of simulated wind errors. These may be determined from analysis of sequences of whole-fire mapping, such as may be obtained from infrared imagery.

3. Production of weather and fire phenomena, including the timing, magnitude, and location, of such effects as fire whirls, pyrocumuli, or horizontal roll vortices. As all of these result from fire–weather interactions, this criterion represents an additional level of verification possible with dynamic models beyond what is possible with kinematic models such as FARSITE, Prometheus, etc. These phenomena may be documented by standard meteorological sensors such as weather radar, or caught by sensors or photographs during research observations, but commonly are anecdotal remarks, video, or photographs by fire suppression personnel or the public, possibly brought to light in incident fatality reports.

These requirements are relevant whether the simulation is to serve as a research tool or in an application – the distinction between which has not historically been made with kinematic fire models, as the modeling procedure has been much the same for either purpose. Going forward, as dynamic models such as CAFMs take a more expanded role in both research and operations, key distinctions and rules

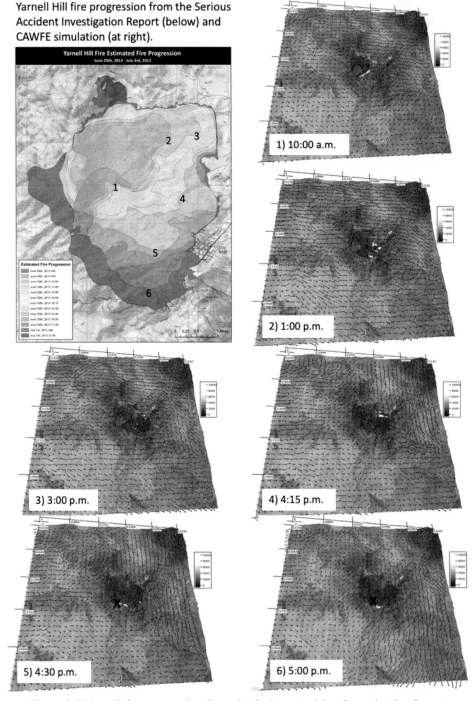

Figure 8.7 Yarnell fire progression from the Serious Accident Investigation Report (top left) and a sequence of times during a CAWFE simulation (numbered 1–6). Reprinted from Coen and Schroeder (2017)

A black and white version of this figure will appear in some formats. For the color version, please refer to the plate section.

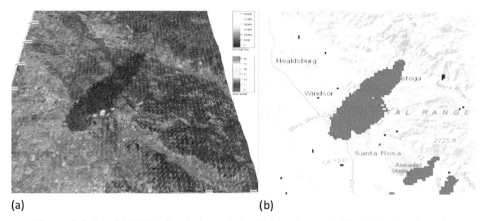

Figure 8.8 (a) CAWFE simulation of the Tubbs fire at 4:09 A.M. showing the downwind run of over 12 miles into the town of Santa Rosa in approximately 3 hours. The protrusion on the lower right side, orthogonal to the wind, occurred as the flank of the fire drew itself up a topographic feature. (b) Visible and Infrared Imaging Radiometer Suite (VIIRS) active fire detections at 3:09 A.M. October 9, 2017. Reprinted from Coen et al. (2018a)
A black and white version of this figure will appear in some formats. For the color version, please refer to the plate section.

should be observed. When asking how good a research simulation is, the criterion should be a test of the physical representation of current understanding as expressed in the model, where a departure of the simulation reflects a gap in understanding. The pernicious expansion of ad-hoc "calibration" throughout the history of fire modeling into the realm of research, whether through recommending spread rate adjustment factors (Finney 1998; Rothermel and Rinehart 1983) to tune the simulation to observed fire spread or adjusting inputs and rerunning simulations until the simulation is consistent with observations (Stratton 2006), has obscured community understanding about the actual skill of kinematic models. Moreover, these habits have carried over to more physically-based CAFMs, for example, with inappropriate retainment of ad hoc adjustment factors developed for heuristic relationships (e.g. Albini and Baughman's (1979) wind adjustment factor) to CAFMs (e.g. Mandel et al. 2011), allowing for arbitrary (and possibly undeclared) altering of rate of spread. The modeling community should declare when calibration has been used in model simulations. While there is a role for using observations to adjust simulations, discussed next, calibration factors in fire behavior rate of spread simulations should not be modified during simulations that are supposedly "validating" a model.

Distinct from pure research simulations, where ad hoc calibration cannot be done while validating the model, applications such as forecasting (notably, modern weather forecasting) avail themselves of techniques that may produce a better forecast by adjusting simulations with observations, including various data assimilation methodologies, nudging, or cycling. Assimilation of data into wildland fire models has some similarities to other problems and unique challenges. It has been treated as a front tracking problem (Filippi et al., 2013; Rochoux et al. 2018), while other approaches have tried to assimilate physical state variables such as temperature (Mandel et al. 2008; Xue et al. 2012); however, as the former note, the errors in temperature of a misplaced fire line are not Gaussian as many approaches assume. Wildfire models and numerical weather prediction simulations of the airflows in which they occur often have nonlinear instabilities and behaviors discouraging use of adjoint methods. Sequential Monte Carlo is another data assimilation technique used for assimilating observations into wildland fire spread. This is a complex approach, as it uses an ensemble-based methodology and implements the Bayesian recursion algorithms directly on highly nonlinear state–space models (Doucet et al. 2001; Niu et al. 2014). Rather than adjusting the interdependent variables of the atmospheric state, Coen and Schroeder (2013) applied the NWP forecasting technique of cycling to coupled weather–fire modeling, using a sequence of simulations reinitialized throughout a fire's lifetime with updated weather and fire location data (in that case, from the Visible and Infrared Imaging Radiometer Suite (VIIRS)) to maintain forecast skill in light of the 1–2 day limit to predictability of fine-scale atmospheric simulations and unpredictable impacts (e.g. suppression and spotting) on fire behavior.

Consider that a CAFM simulation, either for research or application, is performed. Differences between the model simulation and observations arise. How does one identify the source of the error? Is it a poor component of the model? Errors in an input variable? A misinterpretation of the measurement or how it should be compared to a model variable?

First, studies should assess the fidelity of the weather simulation itself, including whether or not the correct flow regime and weather phenomena have been produced, independent of the underlying fire. The weather surrounding notable fire events encompasses regimes that typically are poorly represented by standard meteorological datasets, and that display extreme wind events that are only beginning to be recognized in meteorological literature. Fire weather encompasses some of the least predictable weather scenarios including convective initiation, especially of high-based shallow clouds and their associated gust fronts.

Though researchers often seek rigorous comparisons between model output and exact solutions or observations, this is rarely possible or meaningful. Airflows for which analytical solutions are available for direct comparison are few and typically very idealized. Time series at an individual surface meteorological station in complex terrain and at a model grid point are in general not meaningfully comparable due to the representativeness issues and biases in the station data, and to the interpretation of the modeled grid volume average and temporal variability (often lumped into "gusts," e.g. Coen et al. 2018a). As surface weather stations are sparsely located in the remote regions where wildland fires occur and, particularly at land management agency stations sited in forested terrain, are known to suffer biases, and where low beams of weather radar are often blocked, preventing low level estimates of near surface winds, standard sources of boundary layer weather information are lacking for fire weather applications. In particular, airflows in complex terrain (Whiteman 2000) are extremely important in shaping many of the most destructive wildland fire events. The mathematical theory and simulation of downslope wind events (e.g. Eliassen and Palm 1960; Durran 1990) is very challenging, and a wide variety of downslope wind regimes have been associated with destructive wildland fires. Recent studies have identified gaps in meteorological understanding about the microscale extrema in such events such as the Tubbs Fire and the Camp Fire and the ability of community mesoscale models to represent either the meteorology or the fires that result (Coen et al. 2018a, 2020). When simulated fires, via their fire behavior module, are strongly driven by environmental or fire-induced winds, the result is very sensitive to the magnitude and structure of the wind field, usually more so than to natural variations in other environmental factors.

This may at first appear contradictory, as strong fire-induced winds depend on large and rapid heat releases from the fire – a factor normally tied to fuel properties. While effects of fuel load on ROS and fuel moisture content on ROS may individually be weak on flat ground, each can accelerate fire spread up sloped terrain and their effects can amplify each other on sloped terrain through the connecting medium of the atmosphere (Coen et al. 2018b). Through a "mass fire" effect, heat release creates a strong plume, which draws air in at its base, hence increasing the wind across the fire line and thus increasing the ROS, which increases the amount of fuel on fire, in turn increasing the plume strength and the wind across the fire line, increasing the ROS, and so on, thus capturing the feedback that leads to very large fires or "firestorms." While CAWFE simulations show that relatively small changes in fuels, such as those associated with limited area fuel treatments, have some limited impacts on fires climbing terrain (e.g. the Rim Fire (Tarnay et al. 2020) or the King Fire (Coen et al. 2018b), errors in fuel properties such as canopy fuel loads derived from LANDFIRE can be 50–200%, producing very different fire evolution (Tarnay et al. 2020). The identification of appropriate fuel models from remote sensing data can improve CAFM simulations, notably in recovering disturbed areas (Coen et al. 2018b; Stavros et al. 2018).

In contrast, in wind-driven fires descending slopes, lower surface or canopy fuel moistures (Coen and Schroeder 2015) or higher fuel loads also increase the plume strength but do not draw winds across the fire line, and thus do not increase rate of spread. At extreme heat release levels, plumes may draw winds back against the fire line, opposing fire spread, so that higher fuel loads do not increase the rate of spread in downhill conditions. Therefore, errors in fuel loads in these cases do not necessarily impact the accuracy of fire spread. In summary, the impact of errors in fuels varies, depending on the scenario, but the most significant impact in simulations is on fires climbing terrain, where fuel mapping techniques have led to errors in fuel loads by factors of 1–3.

CFAMS are built from various components, some sophisticated and some highly idealized. The interactions among component sub-models is in the nature of the coupling, and widely accepted components may behave very differently in a coupled framework. A typical landscape model may use the traditional Rothermel approach to fire front propagation to set the heat release rate within the atmospheric boundary layer of the model. So-called errors in components such as the Rothermel prescription have been widely described (it represents steady-state ROS, etc.) but in many simulations the likely primary factors that cause incorrect outcomes are not the ROS model but the wind speeds. When implemented in a coupled framework – particularly in complex terrain – the modeled winds driving the fire include strong spatial and temporal variability of near surface winds and fire-induced winds. An empirical ROS model can serve as an effective parameterization of the relationship between ROS and environmental factors such as wind and slope, since the empirical factors for wind and slope were derived from many fires, in essence building-in some of the intrinsic coupling that occurs in nature.

The point-to-point direct comparison of measured variables – validation as done in experimental building fires, in which one instruments a room then burns it, does not transfer smoothly to outdoor flows embedded in a complex turbulent environment, with variable fuels. For example, surface meteorological station measurements in complex terrain and forested areas are highly suspect. Many issues, including the merits of assumed vertical wind profiles used for vertical interpolation to station height, the meaning of model grid box velocity, the reporting frequency of stations, station bias and error, etc. suggest that a literal direct comparison of model winds interpolated to a surface weather station point is problematic. In airflows in complex terrain this is not a viable approach.

There is the strong potential for misinterpretation or an overly literal interpretation of point-based observations, and, furthermore, we emphasize that the notion of a well-defined fire front spread rate can itself be misleading. In the Tubbs fire, for example, there was a vast production of embers. The evaluation of model performance might compare simulated fire extent to, for instance, satellite observations using active fire detection data (Coen and Schroeder 2013) but do the satellite-based fire detection pixels represent the flaming front (i.e. what the model actually produces), the embers tossed ahead of the fire on the ground, or embers in the smoke cloud, all of which emit strong heat signals detected by the satellite cameras? A high-fidelity representation of the full range of key environmental effects and feedback on fire propagation will require a dedicated combination of measurement, simulation, and operational input to make progress.

8.5 Discussion and Conclusion

Coupled fire–atmosphere feedback is essential for modeling wildland fire spread, especially extreme fire phenomena. The suite of current and emerging tools capable of modeling this complexity now dominate wildland fire research and are starting to be applied to fire operations, training and planning. Some of the barriers to progress and challenges to validating these tools highlighted in this chapter suggest more emphasis on three areas: a scale-dependent and purposeful approach to comparing model results with appropriate observations, recognizing the limitations of each; quantification of the errors and underspecifications in fuel properties and the impact of each; and assessing large-scale simulations and directing observations to address priority research gaps from a position informed by the vast catalog of atmospheric scientific research.

Existing field experimental data sets do not generally contain the full environment description to enable an adequate quantitative assessment of process-based models (Pimont et al. 2017; Hoffman et al. 2018). Moreover, particularly as one moves to larger scale fires, data sets should be constructed to answer pressing research questions, rather than be gathered ad hoc; a hypothesis-driven approach helps to frame this requirement. As the research community moves into developing new experiments to evaluate CFAMs (e.g. Ottmar et al. 2016; Prichard et al. 2019) it is necessary to adopt best practices of atmospheric research experiments for the appropriate scale and to specify the critical input variables and parameters for the evaluation of current and future CFAMs (e.g. Duff et al. 2018).

How will measurements of prescribed fire contribute to improved model subgrid scale parameterizations? These fires constitute a step beyond the laboratory and allow investigators to focus on particular effects such as merging, flanking, and ignition patterns, which should lead to specialized parameterizations. Further experiments on prescribed fires will benefit from more in-depth understanding of the atmospheric boundary layer and more sophisticated measurements (Clements et al. 2008, 2016), along the lines of those used to investigate boundary layer processes (Lenschow et al. 2012). Approaches similar to Pimont et al. (2017), who suggested a rule of thumb to estimate the number of anemometers required in an experimental fire as a function of the height of the canopy and the size of the experimental plot, can be used as a starting point to estimate the sampling requirements for fuel structure measurements, fuel moisture, and fire behavior quantities.

Remote sensing with radar, acoustics and lidar, profilers, and sensor developments for autonomous vehicles have and can be further applied to sampling of the wildland environment, considering that sampling strategies employed on small experimental fires often do not scale to large wildland fires due to their size, potential for rapid expansion on the surface, beam blockage and limited length, and limited access. In pursuing an understanding of convective clouds, atmospheric circulations, and boundary layer processes, to name a few, atmospheric science researchers have employed remote and in situ sensing, theory, and modeling for more than half a century to understand the environment in which fires occur. Much fundamental information about atmospheric circulation and the behavior of fires is available from past atmospheric physical science research concerning how a fluid responds to heat sources. Given this knowledge base, priority fire research questions can be developed, and, from an informed perspective, observational needs to either answer key questions or to test model-originated hypotheses can be designed.

More often than not, an evaluation of a fire spread model, either against a large dataset, a few fire spread data points or other model outputs, is able to show an apparent fit between model and experimental results (Cheney et al. 2012; Menage et al. 2012; Cruz et al. 2013; Dupuy et al. 2014). When considering CFAMs, the large number of model degrees of freedom makes it more likely that an agreement can be artificially forced, at least in simple cases and simple air flows. Revealingly, where strong airflows in complex terrain are incorrect in either direction or flow regime, no amount of adjusting will close the gap. However, the consistency of model parameter choices made to produce an acceptable fit in one case will need to stand up to further comparison in different cases and conditions. It can be contended that a "fitting" approach to model validation is inefficient, counter-productive, and ultimately ineffective from the standpoint of model improvement. Given the typical coarseness and number of unverified assumptions that presently characterize CFAMs, including the sub-grid processes driving fire propagation, and the uncertainty and randomness inherent in any measurement, modelers should strive also to show how the models fail to replicate observed behavior (Watts 1987), or that the methodology used for evaluation might produce counterintuitive results (Gilleland et al. 2009). By investigating a model's unrealistic or erroneous results as well as areas of agreement, the community will contribute to their improvement to a level where they perform adequately over a broad range of conditions.

On longer time scales, in our changing climate, fire occurrence will be modified by expanding fire seasons and new rainfall and drought patterns. Di Virgilio et al. (2019) found increasing risk for severe wildfires in Australia under expected climate change scenarios due to increasing atmospheric instability and dryness events quantified by the C-Haines index. This prospect translates to other regions also impacted by climate change and raises the stakes for improving model simulations of extreme fire behavior. Model improvement will come from, among other endeavors, the expansion of the atmospheric science of wildfire dynamics and further recognition of wildfire as an intrinsic part of weather phenomena.

We note finally that scientists in the structural fire community have many of the same challenges comparing models to data. Structural fires involve the same fluid dynamics, and many of the same environmental effects and fuel conditions, as well as myriad combustion products from flammable building materials. While a structural fire in laboratory conditions is highly controlled and well separated from surrounding environmental effects, in real structural fires, or in urban fire scenarios wind and geometry can have an enormous effect on the rate of spread. To address these challenges this community has produced initial steps toward the "measurement and computation of fire phenomena" (MaCFP) using benchmark experiments, with adequate sampling to test targeted behavior, such as plume pulsation rate in pool fires (Brown et al. 2018). At the initial stage of intercomparison, the focus has been on gas phase burning and simple statistics for basic comparisons. More sophisticated statistical measures of validation are necessary at later stages. Moreover, as described in Chapter 1, pyrolysis and thermokinetics play a key role in fire spread in wildland fire as well as structural fire, and careful intercomparison of quantities such as heat release and mass loss rate from models of surface fire spread (Leventon et al. 2021) will make an important link to the fuels community.

The notion of benchmark experiments, both laboratory and field, is a recurring theme of this chapter, together with model sensitivity studies and new approaches to quantify errors and uncertainty in both models and observations. The role of fire management in the wildland fire community mirrors operational roles in other disciplines; how managers, firefighters, and researchers perceive uncertainty, the different institutional cultures, and the accuracy really required from predictions are all critical areas for which criteria are needed to build confidence in existing and future modeling tools.

References

Agee, JK, Bahro, B, Finney, MA, Omi, PN, Sapsis, DB, Skinner, CN, van Wagtendonk, JW, Weatherspoon, CP (2000) The use of shaded fuel breaks in landscape fire management. *Forest Ecology and Management* **127**(1–3), 55–66.

Agee, JK, Wright, CS, Williamson, N, Huff, MH (2002) Foliar moisture content of Pacific Northwest vegetation and its relation to wildland fire behavior. *Forest Ecology and Management* **167**(1–3), 57–66.

Albini, FA (1976) *Estimating Wildfire Behaviour and Effects*. USDA Forest Service, Intermountain Forest and Range Experimental Station General Technical Report No. INT-30, Ogden, UT.

Albini, FA, Baughman, RG (1979) *Estimating Wind Speeds for Predicting Wildland Fire Behavior*. Research Paper INT-RP-221. Ogden, UT: USDA Forest Service, Intermountain Forest and Range Experiment Station.

Alexander, ME, Cruz, MG (2006) Evaluating a model for predicting active crown fire rate of spread using wildfire observations. *Canadian Journal of Forest Research* **36**(11), 3015–3028.

Alexander, ME, Cruz, MG (2013a) Are the applications of wildland fire behaviour models getting ahead of their evaluation again? *Environmental Modelling & Software* **41**, 65–71.

Alexander, ME, Cruz, MG (2013b) Limitations on the accuracy of model predictions of wildland fire behaviour: a state-of-the-knowledge overview. *The Forestry Chronicle* **89**(3), 372–383.

Alexander, ME, Stefner, CN, Mason, JA, Stocks, BJ, Hartley, GR, Maffey, ME, Wotton, BM, Taylor, SW, Lavoie, N, Dalrymple, GN (2004) *Characterizing the Jack Pine–Black Spruce Fuel Complex of the International Crown Fire Modelling Experiment (ICFME)*. National Resources Canada, Canadian Forest Service, Northern Forestry Centre, Edmonton, AB. Information Report NOR-X-393.

Alexandrov, GA, Ames, D, Bellocchi, G, Bruen, M, Crout, N, Erechtchoukova, M, Hildebrandt, A, Hoffman, F, Jackisch, C, Khaiter, P, Mannina, G, Matsunaga, T, Purucker, ST, Rivington, M, Samaniego, L (2011) Technical assessment and evaluation of environmental models and software: Letter to the Editor. *Environmental Modelling & Software* **26**(3), 328–336.

Andela, N, Morton, DC, Giglio, L, Paugam, R, Chen, Y, Hantson, S, van der Werf, GR, Randerson, JT (2019) The Global Fire Atlas of individual fire size, duration, speed and direction. *Earth System Science Data* **11**, 529–552.

Anderson, HE (1982) *Aids to Determining Fuel Models for Estimating Fire Behavior*. General Technical Report INT-122. USDA Forest Service Intermountain Forest and Range Experiment Station, Ogden, UT.

Aumond, P, Masson, V, Lac, C, Gauvreau, B, Dupont, S, Berengier, M (2013) Including the drag effects of canopies: Real case large-eddy simulation studies. *Boundary-Layer Meteorology* **146**, 65–80.

Bacciu, V, Arca, B, Pellizzaro, G, Salis, M, Ventura, A, Spano, D, Duce, P (2009) Mediterranean maquis fuel model development and mapping to support fire modeling. *EGU General Assembly Conference Abstracts*, vol. 11.

Balbi, J-H, Rossi, J-L, Marcelli, T, Chatelon, F-J (2010) Physical modeling of surface fire under nonparallel wind and slope conditions. *Combustion Science and Technology* **182**(7), 922–939.

Blocken, B, Gualtieri, C (2012) Ten iterative steps for model development and evaluation applied to computational fluid dynamics for environmental fluid mechanics. *Environmental Modelling & Software* **33**, 1–22.

Brown, A, Bruns, M, Gollner, M, Hewson, J, Maragkos, G, Marshall, A, McDermott, R, Merci, B, Rogaume, T, Stoliarov, S, Torero, J, Trouve, A, Wang, Y, Weckman, E (2018) Proceedings of the First Workshop Organized by the IAFSS Working Group on Measurement and Computation of Fire Phenomena (MaCFP). *Fire Safety Journal* **101**, 1–17.

Brown, JK, Bevins, CD (1986) *Surface Fuel Koadings and Predicted Fire Behavior for Vegetation Types in the Northern Rocky Mountains*. USDA Forest Service Research Note INT-358, Intermountain Forest and Range Experiment Station, Ogden UT.

Brown, JK, See, TE (1981) *Downed Dead Woody Fuel and Biomass in the Northern Rocky Mountains.* USDA Forest Service Technical Report, INT-117, Intermountain Forest and Range Experiment Station, Ogden UT.

Burgan, RE (1987) *Concepts and Interpreted Examples in Advanced Fuel Modeling.* USDA For. Serv. Res. Pap. INT-238, Intermountain Forest and Range Experiment Station, Ogden, UT.

Burrows, N, Ward, B, Robinson, A (1991) Fire behaviour in spinifex fuels on the Gibson Desert Nature Reserve, Western Australia. *Journal of Arid Environments* **20**(2), 189–204.

Butler, B, Cohen, J, Latham, D, Schuette, R, Sopko, P, Shannon, K, Jimenez, D, Bradshaw, L (2004) Measurements of radiant emissive power and temperatures in crown fires. *Canadian Journal of Forest Research* **34**, 1577–1587.

Butler, B, Teske, C, Jimenez, D, O'Brien, J, Sopko, P, Wold, C, Vosburgh, M, Hornsby, B, Loudermilk, E (2016) Observations of energy transport and rate of spreads from low-intensity fires in longleaf pine habitat: RxCADRE 2012. *International Journal of Wildland Fire* **25**(1), 76–89.

Canfield, JM, Linn, RR, Sauer, JA, Finney, M, Forthofer, JA (2014) A numerical investigation of the interplay between fireline length, geometry, and rate of spread. *Agricultural and Forest Meteorology* **189–190**, 48–59.

Cardil, A, Monedero, S, Silva, CA, Ramirez, J (2019) Adjusting the rate of spread of fire simulations in real-time. *Ecological Modelling* **395**, 39–44.

Cary, GJ, Keane, RE, Gardner, RH, Lavorel, S, Flannigan, MD, Davies, ID, Li, C, Lenihan, JM, Rupp, TS, Mouillot, F (2006) Comparison of the sensitivity of landscape-fire-succession models to variation in terrain, fuel pattern, climate and weather. *Landscape Ecology* **21**(1), 121–137.

Charland, AM, Clements, CB (2013) Kinematic structure of a wildland fire plume observed by Doppler lidar. *Journal of Geophysical Research: Atmospheres* **118**, 3200–3212.

Cheney, NP, Gould, JS (1995) Fire growth in grassland fuels. *International Journal of Wildland Fire* **5**, 237–247.

Cheney, NP, Gould, JS, Catchpole, WR (1993) The influence of fuel, weather and fire shape variables on fire-spread in grasslands. *International Journal of Wildland Fire* **3**, 31–44.

Cheney, NP, Gould, JS, McCaw, WL, Anderson, WR (2012) Predicting fire behaviour in dry eucalypt forest in southern Australia. *Forest Ecology and Management* **280**, 120–131.

Clark, TL, Coen, JL, Latham, D (2004) Description of a coupled atmosphere–fire model. *International Journal of Wildland Fire* **13**, 49–63.

Clark, TL, Jenkins, MA, Coen, J, Packham, D (1996a) A coupled atmospheric–fire model: Convective feedback on fire line dynamics. *Journal of Applied Meteorology* **35**(6), 875–901.

Clark, TL, Jenkins, MA, Coen, J, Packham, D (1996b) A coupled atmospheric–fire model: Convective Froude number and dynamic fingering. *International Journal of Wildland Fire* **6**(4), 177–190.

Clark TL, Radke, LF, Coen, JL, Middleton, D (1999) Analysis of small-scale convective dynamics in a crown fire using infrared video camera imagery. *Journal of Applied Meteorology* **38**(10), 1401–1420.

Clements, CB (2010) Thermodynamic structure of a grass fire plume. *International Journal of Wildland Fire* **19**(7), 895–902.

Clements, CB, Lareau, NP, Seto, D, Contezac, J, Davis, B, Teske, C, Zajkowski, TJ, Hudak, AT, Bright, BC, Dickinson, MB, Butler, BW, Jimenez, DM, Hiers, JK (2016) Fire weather conditions and fire–atmosphere interactions observed during low-intensity prescribed fires–RxCADRE 2012. *International Journal of Wildland Fire* **25**(1), 90–101.

Clements, CB, Seto, D (2015) Observations of fire–atmosphere interactions and near-surface heat transport on a slope. *Boundary-Layer Meteorology* **154**, 409–426.

Clements, CB, Zhong, S, Bian, X, Heilman, WE, Byun, DW (2008) First observations of turbulence generated by grass fires. *Journal of Geophysical Research: Atmospheres* **113**(D22), D22102.

Clements, CB, Zhong, S, Goodrick, S, Li, J, Potter, BE, Bian, X, Heilman, WE, Charney, JJ, Perna, R, Jang, M, Lee, D, Patel, M, Street, S, Aumann, G (2007) Observing the dynamics of wildland grass fires: FireFlux: A field validation experiment. *Bulletin of the American Meteorological Society* **88**(9), 1369–1382.

Coen, JL (2005) Simulation of the Big Elk Fire using coupled atmosphere–fire modeling. *International Journal of Wildland Fire* **14**(1), 49–59.

Coen, JL (2013) *Modeling Wildland Fires: A Description of the Coupled Atmosphere–Wildland Fire Environment Model (CAWFE)*. NCAR Technical Note NCAR/TN-500+STR.

Coen, JL (2018) Some requirements for simulating wildland fire behavior using insight from coupled weather-wildland fire models. *Fire* **1**(1), 6.

Coen, JL, Cameron, M, Michalakes, J, Patton, EG, Riggan, PJ, Yedinak, KM (2013) WRF-Fire: Coupled weather–wildland fire modeling with the weather research and forecasting model. *Journal of Applied Meteorology and Climatology* **52**(1), 16–38.

Coen, JL, Mahalingam, S, Daily, JW (2004) Infrared imagery of crown-fire dynamics during FROSTFIRE. *Journal of Applied Meteorology* **43**(9), 1241–1259.

Coen, JL, Riggan, PJ (2014) Simulation and thermal imaging of the 2006 Esperanza wildfire in southern California: Application of a coupled weather-wildland fire model. *International Journal of Wildland Fire* **23**(6), 755–770.

Coen, JL, Schroeder, W (2013) Use of spatially refined satellite remote sensing fire detection data to initialize and evaluate coupled weather–wildfire growth model simulations. *Geophysical Research Letters* **40**(20), 5536–5541.

Coen, JL, Schroeder, W (2015) The High Park fire: Coupled weather–wildland fire model simulation of a windstorm-driven wildfire in Colorado's Front Range. *Journal of Geophysical Research: Atmospheres* **120**(1), 131–146.

Coen, JL, Schroeder, W (2017) Coupled weather–fire modeling: From research to operational forecasting. *Fire Management Today* **75**(1), 39–45.

Coen, JL, Schroeder, W, Conway, S, Tarnay, L (2020) Computational modeling of extreme wildland fire events: a synthesis of scientific understanding with applications to forecasting, land management, and firefighter safety. *Journal of Computational Science* **45**, 101152.

Coen, JL, Schroeder, W, Quayle, B (2018a) The generation and forecast of extreme winds during the origin and progression of the 2017 Tubbs Fire. *Atmosphere* **9**(12), 462.

Coen, JL, Stavros, EN, Fites-Kaufman, JA (2018b) Deconstructing the King megafire. *Ecological Applications* **28**(6), 1565–1580.

Countryman, CM (1969) *Project Flambeau ... An Investigation of Mass Fire (1964–1967), final report volume I*. Prepared for Office of Civil Defense under OCD Work Order No. OCD-PS-65-26. Berkeley, CA: USDA Forest Service, Pacific Southwest Forest and Range Experiment Station.

Cruz, MG, Alexander, ME (2013) Uncertainty associated with model predictions of surface and crown fire rates of spread. *Environmental Modelling & Software* **47**, 16–28.

Cruz, MG, Alexander ME, Sullivan, AL (2017) Mantras of wildland fire behaviour modelling: Facts or fallacies? *International Journal of Wildland Fire* **26**(11), 973–981.

Cruz, MG, Gould, JS, Kidnie, S, Bessell, R, Nichols, D, Slijepcevic, A (2015) Effects of curing on grass fires: II. Effect of grass senescence on the rate of fire spread. *International Journal of Wildland Fire* **24**(6), 838–848.

Cruz, MG, McCaw, WL, Anderson, WR, Gould, JS (2013) Fire behaviour modelling in semi-arid mallee-heath shrublands of southern Australia. *Environmental Modelling & Software* **40**, 21–34.

Cruz, MG, Sullivan, AL, Gould, JS, Sims, NC, Bannister, AJ, Hollis, JJ, Hurley, RJ (2012) Anatomy of a catastrophic wildfire: The Black Saturday Kilmore East fire in Victoria, Australia. *Forest Ecology and Management* **284**, 269–285.

Davis, C, Brown, B, Bullock, R (2006) Object-based verification of precipitation forecasts. Part I: Methodology and application to mesoscale rain areas. *Monthly Weather Review* **134**(7), 1772–1784.

Dennison, PE, Brewer, SC, Arnold, JD, Moritz, MA (2014) Large wildfire trends in the western United States, 1984–2011. *Geophysical Research Letters* **41**(8), 2928–2933.

Di Virgilio, G, Evans, JP, Blake, SAP, Armstrong, M, Dowdy, AJ, Sharples, J, McRae, R (2019) Climate change increases the potential for extreme wildfires. *Geophysical Research Letters* **46**(14), 8517–8526.

Doucet, A, Freitas, N, Gordon, N (2001) *Sequential Monte Carlo Methods in Practice*. New York: Springer.

Duff, TJ, Cawson, JG, Cirulis, B, Nyman, P, Sheridan, GJ, Tolhurst, KG (2018) Conditional performance evaluation: using wildfire observations for systematic fire simulator development. *Forests* **9**(4), 189.

Dupuy, JL, Morvan, D (2005) Numerical study of a crown fire spreading toward a fuel break using a multiphase physical model. *International Journal of Wildland Fire* **14**(2), 141–151.

Dupuy, J, Pimont, F, Linn, R, Clements, C (2014) FIRETEC evaluation against the FireFlux experiment: preliminary results. In: Viegas, DX, ed. *Advances in Forest Fire Research*. Coimbra, Portugal: University of Coimbra, pp. 261–274.

Durran, DR (1990) Mountain waves and downslope winds. In: Blumen, W, ed. *Atmospheric Processes over Complex Terrain*, Meteorological Monographs series, vol. 23. Boston, MA: American Meteorological Society, pp. 59–83.

Eliassen, A, Palm, E (1960) On the transfer of energy in stationary mountain waves. *Geofysiske Publikasjoner* **22**(3), 1–23.

Fahnestock, GR, Key, WK (1971) Weight of brushy forest fire fuels from photographs. *Forest Science* **17**(1), 119–124.

Fernandes, PM, Catchpole, WR, Rego, FC (2000) Shrubland fire behaviour modelling with microplot data. *Canadian Journal of Forest Research* **30**(6), 889–899.

Filippi, J-B, Bosseur, F, Mari, C, Lac, C (2018) Simulation of a large wildfire in a coupled fire-atmosphere model. *Atmosphere* **9**(6), 218.

Filippi, J-B, Bosseur, F, Mari, C, Lac, C, Le Moigne, P, Cuenot, B, Veynante, D, Cariolle, D, Balbi, JH (2009) Coupled atmosphere-wildland fire modelling. *Journal of Advances in Modeling Earth Systems* **1**(4), 1–9.

Filippi, J-B, Mallet, V, Nader, B (2014) Representation and evaluation of wildfire propagation simulations. *International Journal of Wildland Fire* **23**(1), 46–57.

Filippi, J-B, Pialat, X, Clements, CB (2013) Assessment of ForeFire/Meso-NH for wildland fire/atmosphere coupled simulation of the FireFlux experiment. *Proceedings of the Combustion Institute* **34**(2), 2633–2640.

Finney, MA (1998) *FARSITE: Fire Area Simulator: Model Development and Evaluation*. US Department of Agriculture Forest Service: Ogden, UT, USA.

Fons, WL (1946) Analysis of fire spread in light forest fuels. *Journal of Agricultural Research* **72**(3), 93–121.

Frankman, D, Webb, BW, Butler, BW, Jimenez, D, Forthofer, JM, Sopko, P, Shannon, KS, Hiers, JK, Ottmar, RD (2013) Measurements of convective and radiative heating in wildland fires. *International Journal of Wildland Fire* **22**(2), 157–167.

Fujioka, FM (2002) A new method for the analysis of fire spread modeling errors. *International Journal of Wildland Fire* **11**(3–4), 193–203.

de Gennaro, M, Billaud, Y, Pizzo, Y, Garivait, S, Loraud, JC, El Hajj, M, Porterie, B (2017) Real-time wildland fire spread modeling using tabulated flame properties. *Fire Safety Journal* **91**, 872–881.

Gilleland, E, Ahijevych, D, Brown, BG, Casati, B, Ebert, EE (2009) Intercomparison of spatial forecast verification methods. *Weather Forecasting* **24**(5), 1416–1430.

Gould, JS, McCaw, WL, Cheney, NP, Ellis, PF, Knight, IK, Sullivan, AL (2008) *Project Vesta: Fire in Dry Eucalypt Forest: Fuel Structure, Fuel Dynamics and Fire Behaviour*. Canberra, ACT: CSIRO Publishing.

Habeeb, RL, Trebilco, J, Wotherspoon, S, Johnson, CR (2005) Determining natural scales of ecological systems. *Ecological Monographs* **75**(4), 467–487.

Hardy, C, Heilman, W, Weise, D, Goodrick, S, Ottmar, R (2008) *Fire Behavior Advancement Plan; a Plan for Addressing Physical Fire Processes within the Core Fire Science Portfolio*. Final report to the Joint Fire Sciences Program Board of Governors.

Hiers, JK, O'Brien, JJ, Mitchell, RJ, Grego, JM, Loudermilk, EL (2009) The wildland fuel cell concept: An approach to characterize fine-scale variation in fuels and fire in frequently burned longleaf pine forests. *International Journal of Wildland Fire* **18**(3), 315–325.

Hoffman, CM, Morgan, P, Mell, W, Parsons, R, Strand, E, Cook, S (2013) Surface fire intensity influences simulated crown fire behavior in lodgepole pine forests with recent mountain pine beetle-caused tree mortality. *Forest Science* **59**(4), 390–399.

Hoffman, CM, Sieg, CH, Linn, RR, Mell, W, Parsons, RA, Ziegler, JP, Hiers, JK (2018) Advancing the science of wildland fire dynamics using process-based models. *Fire* **1**(2), 32.

Hornby, LG (1936) Fire control planning in the northern Rocky Mountain region. Progress Report No. 1. Missoula, MT: US Department of Agriculture, Forest Service, Northern Rocky Mountain Forest and Range Experiment Station.

Jakeman, AJ, Letcher, RA, Norton, JP (2006) Ten iterative steps in development and evaluation of environmental models. *Environmental Modelling & Software* **21**(5), 602–614.

Johnson, EA, Miyanishi, K, Weir, JMH (1998) Wildfires in the western Canadian boreal forest: Landscape patterns and ecosystem management. *Journal of Vegetation Science* **9**(4), 603–610.

Kalabokidis, K, Omi, P (1992) Quadrat analysis of wildland fuel spatial variability. *International Journal of Wildland Fire* **2**(4), 145–152.

Keane, RE (2008) Biophysical controls on surface fuel litterfall and decomposition in the northern Rocky Mountains, USA. *Canadian Journal of Forest Research* **38**(6), 1431–1445.

Keane, RE (2013) Describing wildland surface fuel loading for fire management: A review of approaches, methods and systems. *International Journal of Wildland Fire* **22**(1), 51–62.

Keane, RE, Burgan, R, van Wagtendonk, J (2001) Mapping wildland fuels for fire management across multiple scales: integrating remote sensing, GIS, and biophysical modeling. *International Journal of Wildland Fire* **10**(4), 301–319.

Keane, RE, Gray, K, Bacciu, V, Leirfallom, S (2012) Spatial scaling of wildland fuels for six forest and rangeland ecosystems of the northern Rocky Mountains, USA. *Landscape Ecology* **27**(8), 1213–1234.

Keane, RE, Reeves, M (2012) Use of expert knowledge to develop fuel maps for wildland fire management. In: Perera, AH, Drew, C, Johnson, CJ, eds. *Expert Knowledge and Its Application in Landscape Ecology*. New York: Springer, pp. 211–228.

King, KJ, Bradstock, RA, Cary, GJ, Chapman, J, Marsden-Smedley, JB (2008) The relative importance of fine-scale fuel mosaics on reducing fire risk in south-west Tasmania, Australia. *International Journal of Wildland Fire* **17**(3), 421–430.

Lagouvardos, K, Kotroni, V, Giannaros, TM, Dafis, S (2019) Meteorological conditions conducive to the rapid spread of the deadly wildfire in eastern Attica, Greece. *Bulletin of the American Meteorological Society* **100**(11), 2137–2145.

Lenschow, DH, Lothon, M, Mayor, SD, Sullivan, PP, Canut, G (2012) A comparison of higher-order vertical velocity moments in the convective boundary layer from lidar with in situ measurements and LES. *Boundary-Layer Meteorology* **143**, 107–123.

Leventon, I, Batiot, B, Bruns, M, Hostikka, S, Nakamura, Y, Reszka, P, Rogaume, T, Stoliarov, S (2021) *The MaCFP Condensed Phase Working Group: A Structured, Global Effort towards Pyrolysis Model Development*, ASTM Selected Technical Papers (STP), Atlanta, GA, US [online]. Available from https://tsapps.nist.gov/publication/get_pdf.cfm?pub_id=933681 (last accessed November 26, 2021).

Linn, RR (1997) *A Transport Model for Prediction of Wildfire Behavior*. PhD Thesis, New Mexico State University, Los Alamos National Laboratory, Scientific Report LA13334-T.

Linn, RR, Anderson, K, Winterkamp, J, Brooks, A, Wotton, M, Dupuy, JL, Wotton, M, Edminster, C (2012) Incorporating field wind data into FIRETEC simulations of the International Crown Fire Modeling Experiment (ICFME): Preliminary lessons learned. *Canadian Journal of Forest Research* **42**(5), 879–898.

Linn, RR, Cunningham, P (2005) Numerical simulations of grass fires using a coupled atmosphere–fire model: Basic fire behavior and dependence on wind speed. *Journal of Geophysical Research-Atmospheres* **110**(D13), D13107.

Linn, RR, Goodrick, SL, Brambilla, S, Brown, MJ, Middleton, RS, O'Brien, JJ, Hiers, JK (2020) QUIC-fire: A fast-running simulation tool for prescribed fire planning. *Environmental Modelling & Software* **125**, 104616.

Linn, RR, Reisner, J, Colman, JJ, Winterkamp, J (2002) Studying wildfire behavior using FIRETEC. *International Journal of Wildland Fire* **11**(3–4), 233–246.

Linn, RR, Winterkamp, J, Colman, J, Edminster, C, Bailey, JD (2005) Modeling interactions between fire and atmosphere in discrete element fuel beds. *International Journal of Wildland Fire* **14**(1) 37–48.

Liu, Y, Kochanski, A, Baker, KR, Mell, W, Linn, R, Paugam, R, Mandel, J, Fournier, A, Jenkins, M-A, Goodrick, S, Achtemeier, G, Zhao, F, Ottmar, R, French, NHF, Larkin, N, Brown, T, Hudak, A, Dickinson, M, Potter, B, Clements, C, Urbanski, S, Prichard, S, Watts, A, McNamara, D (2019) Fire behaviour and smoke modelling: model improvement and measurement needs for next-generation smoke research and forecasting systems. *International Journal of Wildland Fire* **28**(8), 570–588.

Lopes, AMG, Sousa, ACM, Viegas, DX (1995) Numerical simulation of turbulent flow and fire propagation in complex topography. *Numerical Heat Transfer, Part A: Applications* **27**(2), 229–253.

Lorenz, EN (1969) The predictability of a flow which possesses many scales of motion. *Tellus* **21**(3), 289–307.

Loudermilk, EL, O'Brien, J, Mitchell, RJ, Hiers, JK, Cropper, WP, Grunwald, S, Grego, J, Fernandez, J (2012) Linking complex forest fuel structure and fire behaviour at fine scales. *International Journal of Wildland Fire* **21**(7), 882–893.

Mandel, J, Beezley, J, Coen, J, Kim, M (2009) Data assimilation for wildland fires: Ensemble Kalman Filters in coupled atmosphere–surface models. *IEEE Control Systems Magazine* **29**(3), 47–65.

Mandel, J, Beezley, JD, Kochanski, AK (2011) Coupled atmosphere–wildland fire modeling with WRF-Fire version 3.3. *Geoscientific Model Development* **4**, 591–610.

Mandel, J, Bennethum, LS, Beezley, JD, Coen, JL, Douglas, CC, Kim, M, Vodacek, A (2008) A wildland fire model with data assimilation. *Mathematics and Computers in Simulation* **79**(3), 584–606.

Marino, E, Dupuy, JL, Pimont, F, Guijarro, M, Hernando, C, Linn, R (2012) Fuel bulk density and fuel moisture content effects on fire rate of spread: A comparison between FIRETEC model predictions and experimental results in shrub fuels. *Journal of Fire Sciences* **30**(4), 277–299.

McArthur, AG (1967) *Fire Behaviour in Eucalypt Forests*. Commonwealth of Australia, Forestry and Timber Bureau No. Leaflet 107, Canberra, ACT.

McCaw, WL, Gould, JS, Cheney, NP, Ellis, PFM, Anderson, WR (2012) Changes in behaviour of fire in dry eucalypt forest as fuel increases with age. *Forest Ecology and Management* **271**, 170–181.

Mell, WE, Jenkins, MA, Gould, J, Cheney, P (2007) A physics-based approach to modelling grassland fires. *International Journal of Wildland Fire* **16**(1), 1–22.

Mell, WE, McGrattan, KB, Baum, HR (1996) Numerical simulation of combustion in fire plumes. *Symposium (International) on Combustion*, **26**(1), 1523–1530.

Menage, D, Chetehouna, K, Mell, W (2012) Numerical simulations of fire spread in a Pinus pinaster needles fuel bed. *Journal of Physics: Conference Series* **395**(1), 012011.

Moinuddin, KAM, Sutherland, D, Mell, W (2018) Simulation study of grass fire using a physics-based model: Striving towards numerical rigour and the effect of grass height on the rate of spread. *International Journal of Wildland Fire* **27**(12), 800–814.

Morandini, F, Simeoni, A, Santoni, PA, Balbi, J-H (2005) A model for the spread of fire across a fuel bed incorporating the effects of wind and slope. *Combustion Science and Technology* **177**(7), 1381–1418.

Morvan, D, Dupuy, J (2001) Modeling of fire spread through a forest fuel bed using a multiphase formulation. *Combustion and Flame* **127**(1–2), 1981–1994.

Mukherjee, S, Schalkwuk, J, Jonker, HJJ (2016) Predictability of dry convective boundary layers: An LES study. *Journal of the Atmospheric Science* **73**(7), 2715–2727.

Niu, S, Luo, Y, Dietze, M, Keenan, TF, Shi, Z, Li, J, Chapin III, FS (2014) The role of data assimilation in predictive ecology. *Ecosphere* **5**(5), 1–16.

O'Brien, JJ, Loudermilk, EL, Hornsby, B, Hudak, AT, Bright, BC, Dickinson, MB, Hiers, JK, Teske, C, Ottmar, RD (2016) High-resolution infrared thermography for capturing wildland fire behaviour: RxCADRE 2012. *International Journal of Wildland Fire* **25**(1), 62–75.

Ottmar, RD, Hiers, JK, Butler, BW, Clements, CB, Dickinson, MB, Hudak, AT, O'Brien, JO, Potter, BE, Rowell, EM, Strand, TM, Zajkowski, TJ (2016) Measurements, datasets and preliminary results from the RxCADRE project–2008, 2011 and 2012. *International Journal of Wildland Fire* **25**(1), 1–9.

Ottmar, RD, Sandberg, DV, Riccardi, CL, Prichard, SJ (2007) An overview of the fuel characteristic classification system: Quantifying, classifying, and creating fuelbed for resource planning. *Canadian Journal of Forest Research* **37**(12), 2383–2393.

Parsons, RA, Mell, WE, McCauley, P (2011) Linking 3D spatial models of fuels and fire: Effects of spatial heterogeneity on fire behavior. *Ecological Modelling* **222**(3), 679–691.

Parsons, RA, Pimont, F, Wells, L, Cohn, G, Jolly, WM, de Coligny, F, Rigolet, E, Dupuy, J-L, Mell, W, Linn, RR (2018) Modeling thinning effects on fire behavior with STANDFIRE. *Annals of Forest Science* **75**(1), 7.

Peace, M, Mattner, T, Mills, G, Kepert, J, McCaw, L (2015) Fire-modified meteorology in a coupled fire–atmosphere model. *Journal of Applied Meteorology and Climatology*, **54**(3), 704–720.

Pimont, F, Dupuy, JL, Linn, RR, Dupont, S (2009) Validation of FIRETEC wind-flows over a canopy and a fuel-break. *International Journal of Wildland Fire* **18**(7), 775–790.

Pimont, F, Dupuy, J-L, Linn, RR, Dupont, S (2011) Impacts of tree canopy structure on wind flows and fire propagation simulated with FIRETEC. *Annals of Forest Science* **68**, 523–530.

Pimont, F, Dupuy, JL, Linn, RR, Parsons, R, Martin-StPaul, N (2017) Representativeness of wind measurements in fire experiments: Lessons learned from large-eddy simulations in a homogeneous forest. *Agricultural and Forest Meteorology* **232**, 479–488.

Prichard, S, Larkin, NS, Ottmar, R, French, NHF, Baker, K, Brown, T, Clements, C, Dickinson, M, Hudak, A, Kochanski, A, Linn, R, Liu, Y, Potter, B, Mell, W, Tanzer, D, Urbanski, S, Watts, A (2019) The fire and smoke model evaluation experiment: A plan for integrated, large fire–atmosphere field campaigns. *Atmosphere* **10**(2), 66.

Rabin, SS, Melton, JR, Lasslop, G, Bachelet, D, Forrest, M, Hantson, S, Kaplan, JO, Li, F, Mangeon, S, Ward, DS, Yue, C, Arora, VK, Hickler, T, Kloster, S, Knorr, W, Nieradzik, L, Spessa, A, Folberth, GA, Sheehan, T, Voulgarakis, A, Kelley, DI, Prentice, IC, Sitch, S, Harrison, S, Arneth, A (2017) The Fire Modeling Intercomparison Project (FireMIP), phase 1: Experimental and analytical protocols with detailed model descriptions. *Geoscientific Model Development* **10**, 1175–1197.

Radke, LR, Clark, TL, Coen, JL, Walther, C, Lockwood, RN, Riggin, PJ, Brass, J, Higgins, R (2000) The WildFire Experiment (WiFE): Observations with airborne remote sensors. *Canadian Journal of Remote Sensing* **26**(5), 406–417.

Reeves, MC, Ryan, KC, Rollins, MG, Thompson, TG (2009) Spatial fuel data products of the LANDFIRE project. *International Journal of Wildland Fire* **18**(3), 250–267.

Reich, RM, Lundquist, JE, Bravo, VA (2004) Spatial models for estimating fuel loads in the Black Hills, South Dakota, USA. *International Journal of Wildland Fire* **13**(1), 119–129.

Riccardi, CL, Ottmar, RD, Sandberg, DV, Andreu, A, Elman, E, Kopper, K, Long, J (2007) The fuelbed: A key element of the Fuel Characteristic Classification System. *Canadian Journal of Forest Research* **37**(12), 2394–2412.

Rochoux, MC, Collin, A, Zhang, C, Trouvé, A, Lucor, D, Moireau, P (2018) Front shape similarity measure for shape-oriented sensitivity analysis and data assimilation for Eikonal equation. *ESAIM: Proceedings and Surveys* **63**(ESAIM: ProcS), 258–279.

Rocca, ME (2009) Fine-scale patchiness in fuel load can influence initial post-fire understory composition in a mixed conifer forest, Sequoia National Park, California. *Natural Areas Journal* **29**(2), 126–133.

Rothermel, RC (1972) *A Mathematical Model for Predicting Fire Spread in Wildland Fuels*. USDA Forest Service No. Research Paper INT-115, Ogden, UT.

Rothermel, RC, Rinehart, GC (1983) *Field Procedures for Verification and Adjustment of Fire Behavior Predictions*. Research report. Department of Agriculture, Forest Service, Intermountain Forest and Range Experiment Station: Ogden, UT.

Rykiel, EJ (1996) Testing ecological models: the meaning of validation. *Ecological Modelling* **90**(3), 229–244.

Sandberg, DV, Ottmar, RD, Cushon, GH (2001) Characterizing fuels in the 21st century. *International Journal of Wildland Fire* **10**(3–4), 381–387.

Sandberg, DV, Riccardi, CL, Schaaf, MD (2007) Reformulation of Rothermel's wildland fire behaviour model for heterogeneous fuelbeds. *Canadian Journal of Forestry Research* **37**(12), 2438–2455.

Santoni, P-A, Filippi, J-B, Balbi, J-H, Bosseur, F (2011) Wildland fire behaviour case studies and fuel models for landscape-scale fire modeling. *Journal of Combustion* Article ID 613424.

Schag, GM, Stow, DA, Riggan, PJ, Tissell, RG, Coen, JL (2021) Examining landscape-scale fuel and terrain controls of wildfire spread rates using repetitive airborne thermal infrared (ATIR) imagery. *Fire* **4**(1), 6.

Schroeder, W, Oliva, P, Giglio, L, Csiszar, I (2014) The new VIIRS 375 m active fire detection data product: Algorithm description and initial assessment. *Remote Sensing of Environment* **143**, 85–96.

Scott, JH, Burgan, RE (2005) *Standard Fire Behavior Fuel Models: A Comprehensive Set for Use with Rothermel's Surface Fire Spread Model*. General Technical Report RMRS-GTR-153. Fort Collins, CO: U.S. Department of Agriculture, Forest Service, Rocky Mountain Research Station.

Simpson, CC, Sharples, JJ, Evans, JP (2016) Sensitivity of atypical lateral fire spread to wind and slope. *Geophysical Research Letters* **43**(4), 1744–1751.

Stavros, EN, Coen, J, Peterson, B, Singh, H, Kennedy, K, Ramirez, C, Schimel, D (2018) Use of imaging spectroscopy and LIDAR to characterize fuels for fire behavior prediction. *Remote Sensing Applications: Society and Environment* **11**, 41–50.

Stephens, SL, Lydersen, JM, Collins, BM, Fry, DL, Meyer, MD (2015) Historical and current landscape-scale ponderosa pine and mixed conifer forest structure in the Southern Sierra Nevada. *Ecosphere* **6**(5), 1–63.

Stocks, BJ (1987) Fire behavior in immature jack pine. *Canadian Journal of Forest Research* **17**(1), 80–86.

Stocks, BJ, Alexander, ME, Wotton, BM, Stefner, CN, Flannigan, MD, Taylor, SW, Lavoie, N, Mason, JA, Hartley, GR, Maffey, ME, Dalrymple, GN, Blake, TW, Cruz, MG, Lanoville, RA (2004) Crown fire behaviour in a northern jack pine – Black spruce forest. *Canadian Journal of Forest Research* **34**(8), 1548–1560.

Stow, D, Riggan, P, Schag, G, Brewer, W, Tissell, R, Coen, J, Storey, E (2019) Assessing uncertainty and demonstrating potential for estimating fire rate of spread at landscape scales based on time sequential airborne thermal infrared imaging. *International Journal of Remote Sensing* **40**(13), 4876–4897.

Stratton, RD (2006) *Guidance on Spatial Wildland Fire Analysis: Models, Tools, and Techniques*. Research report. Department of Agriculture, Forest Service, Rocky Mountain Research Station: Fort Collins, CO.

Sullivan, AL (2017) Inside the inferno: Fundamental processes of wildland fire behaviour part 1: Combustion chemistry and heat release. *Current Forestry Reports* **3**, 132–149.

Sun, R, Krueger, SK, Jenkins, MA, Zulauf, MA, Charney, JJ (2009) The importance of fire–atmosphere coupling and boundary-layer turbulence to wildfire spread. *International Journal of Wildland Fire* **18**(1), 50–60.

Tarnay, L, Coen, J, Kennedy, K, McElhaney, M, Evans, K, Ramirez, C (2020) Modeled effects of fuel reduction on rim fire daily smoke emissions. Virtual Conference. *International Association of Wildland Fire 3rd International Smoke Symposium ISS3#*, April, Rayleigh, NC and Davis, CA.

Taylor, SW, Wotton, BM, Alexander, ME, Dalrymple, GN (2004) Variation in wind and crown fire behaviour in a northern jack pine black spruce forest. *Canadian Journal of Forest Research* **34**(8), 1561–1576.

Thurston, W, Fawcett, RJ, Tory, KJ, Kepert, JD (2016) Simulating boundary-layer rolls with a numerical weather prediction model. *Quarterly Journal of the Royal Meteorological Society* **142**(694), 211–223.

Turner, MG, Romme, WH (1994) Landscape dynamics in crown fire ecosystems. *Landscape Ecology* **9**, 59–77.

Van Wagdendonk, JW, Benedict, JM, Sydoriak, WM (1996) Physical properties of woody fuel particles of Sierra Nevada conifers. *International Journal of Wildland Fire* **6**(3), 117–123.

Van Wagdendonk, JW, Sydoriak, WM, Benedict, JM (1998) Heat content variation of Sierra Nevada conifers. *International Journal of Wildland Fire* **8**(3), 147–158.

Watts, JM (1987) Validating fire models. *Fire Technology* **23**, 93–94.

Whiteman, CD (2000) *Mountain Meteorology: Fundamentals and Applications*. New York: Oxford University Press.

Wotton, BM, Gould, JS, McCaw, WL, Cheney, NP, Taylor, SW (2012) Flame temperature and residence time of fires in dry eucalypt forest. *International Journal of Wildland Fire* **21**(3), 270–281.

Xue, HD, Gu, F, Hu, XL (2012) Data assimilation using sequential Monte Carlo methods in wildfire spread simulation. *ACM Transactions on Modeling and Computer Simulation* **22**(4), 1–25.

Zhang, C, Collin, A, Moireau, P, Trouvé, A, Rochoux, MC (2019) State-parameter estimation approach for data-driven wildland fire spread modeling: Application to the 2012 RxCADRE S5 field-scale experiment. *Fire Safety Journal* **105**, 286–299.

Zhang, C, Rochoux, MC, Tang, W, Gollner, M, Filippi, JB, Trouvé, A (2017) Evaluation of a data-driven wildland fire spread forecast model with spatially-distributed parameter estimation in simulations of the Fireflux field-scale experiment. *Fire Safety Journal* **91**, 758–767.

Zhou X, Mahalingam S, Weise D (2007) Experimental study and large eddy simulation of effect of terrain slope on marginal burning in shrub fuel beds. *Proceedings of the Combustion Institute* **31**(2), 2547–2555.

Ziegler, JP, Hoffman, C, Battaglia, M, Mell, W (2017) Spatially explicit measurements of forest structure and fire behavior following restoration treatments in dry forests. *Forest Ecology and Management* **386**, 1–12.

Index

Printed in the United States
by Baker & Taylor Publisher Services